Letters from the Ecotone

Letters from the Ecotone

Ecology, Theology, and Climate Change

ANDREW NAGY-BENSON
ANDREA LLOYD

RESOURCE *Publications* • Eugene, Oregon

LETTERS FROM THE ECOTONE
Ecology, Theology, and Climate Change

Copyright © 2022 Andrew Nagy-Benson and Andrea Lloyd. All rights reserved. Except for brief quotations in critical publications or reviews, no part of this book may be reproduced in any manner without prior written permission from the publisher. Write: Permissions, Wipf and Stock Publishers, 199 W. 8th Ave., Suite 3, Eugene, OR 97401.

Resource Publications
An Imprint of Wipf and Stock Publishers
199 W. 8th Ave., Suite 3
Eugene, OR 97401

www.wipfandstock.com

PAPERBACK ISBN: 978-1-6667-5831-3
HARDCOVER ISBN: 978-1-6667-5832-0
EBOOK ISBN: 978-1-6667-5833-7

11/29/22

Cover photo, "Addison, Vermont," by Caleb Kenna, copyright © 2022 by Caleb Kenna. Used with permission.

Except where otherwise indicated, scripture quotations are from New Revised Standard Version Bible, copyright © 1989 National Council of Churches in the United States of America (http://www.nrsvbibles.org). Used by permission. All rights reserved worldwide.

On page 53-56, "Principles of Environmental Justice" reprinted, by permission, from The First National People of Color Environmental Leadership Summit, Principles of Environmental Justice, October 1991, Charles Lee, ed., 1992. Copyright © 1991 by United Church of Christ Commission for Racial Justice: New York. (Now Justice and Witness Ministries, a Covenanted Ministry of the United Church of Christ, Cleveland, Ohio).

On page 59, lyrics from "Go Up to the Mountain" by the Monks of the Weston Priory. Copyright © 1978 by The Benedictine Foundation of the State of Vermont, Inc. Reproduced by permission of The Benedictine Foundation of the State of Vermont, Inc.

On page 90-91, "YECA supports Environmental Justice for All Act," from the website of Young Evangelics for Climate Action. Copyright © 2021 by YECA. Reproduced by permission of Young Evangelicals for Climate Action.

On page 126, "blessing the boats" from *How To Carry Water: Selected Poems of Lucille Clifton*. Copyright © 1991 by Lucille Clifton. Reprinted with the permission of The Permissions Company, LLC on behalf of BOA Editions, LTD., boaeditions.org.

To Gwen, Ella, Mary, and Rachael
-Andy Nagy-Benson

To my mother, Jacqueline Roe Lloyd,
from whom I first learned to fall in love
with green, growing things
-Andi Lloyd

Contents

Preface | ix
Acknowledgments | xiii

June 4, 2020 | 1
July 27, 2020 | 2
August 21, 2020 | 8
August 26, 2020 | 12
September 7, 2020 | 18
On Easter Evening | 21
September 13, 2020 | 22
September 21, 2020 | 28
September 23, 2020 | 31
September 30, 2020 | 36
October 20, 2020 | 42
October 28, 2020 | 46
November 18, 2020 | 52

December 26, 2020 | 59
January 14, 2021 | 67
January 31, 2021 | 72
February 24, 2021 | 77
March 5, 2021 | 82
March 28, 2021 | 88
April 8, 2021 | 93
April 28, 2021 | 100
May 6, 2021 | 105
May 19, 2021 | 111
June 4, 2021 | 116
June 23, 2021 | 123
June 7, 2021 | 128

Bibliography | 129

Preface

Dear Reader,

On a sunny Sunday afternoon in October 2017, I slipped into one of the back pews of a church in a tiny town in the Green Mountains of Vermont. I was there to hear a pastor give the annual Rabbi Reichert Bible Talk. And I was there because I'd decided it was time to stop running from a lifelong belief in God, and the talk seemed a good opportunity to strike up a conversation with a clergyperson about how one does that. In many ways, this book began that day—with an afternoon conversation between a pastor and an ecologist.

I was, until recently, a biology professor at Middlebury College in Vermont. For two decades, I taught about climate change and studied its effects on the forests of Alaska and Siberia. The pastor I approached that day is Rev. Andy Nagy-Benson. He has been an ordained minister in the United Church of Christ for over twenty years and is currently the senior pastor at the Congregational Church of Middlebury (UCC).

Our conversation that October afternoon led to another over breakfast a week later. Andy convinced me to give church a try. And I did, the following Sunday. I've gone back just about every Sunday since. The two of us began meeting regularly, Andy having generously offered to help fill the gaping holes in my understanding of Christianity. Every couple of weeks, we'd sit across from each other at a book-strewn table in a church classroom and talk theology. Ecology found its way into those conversations, too. We learned early on that we share a common concern for the well-being of life on Earth. These days, that means we also share a deep concern about climate change. We began to dream about where that shared interest and our growing realization of how much ecology and theology have to say to each other might take us.

My return to church certainly took me farther than I ever imagined it would. Within a year, I had resigned my tenured faculty position and made plans to attend Yale Divinity School, pursuing a call to ordained ministry. Andy and I, meanwhile, took our conversation about ecology, theology, and climate change into the classroom: in January 2019, we team-taught a course at Middlebury College on "Hope in a Time of Climate Change."

The class was designed around the questions that had emerged over the course of our conversations. What might a dialogue between these two long-separated disciplines—ecology and theology—teach us? Where do they speak in harmony? And where do they not? How might ecology and theology together help us to imagine a different future? Our students met the project of intellectual bridge-building with enthusiasm and creativity. They were captivated by the idea that science and religion, which many of them assumed to be adversaries, could work together toward the common goal of flourishing life on Earth. We learned at least as much from them as they did from us. We designed a new class a year later, organized around core concepts that we thought might have resonance in both ecology and theology. Words from which bridges could be built, we thought.

The pandemic put our conversations on hold for a while. And then, one day in July of 2020, when Vermont's Covid restrictions had lifted enough to allow us to get together, we went for a long walk. And we talked. We talked about class. About our lives in a time of pandemic. About climate change. About how to keep this conversation going. Maybe we could go old-school, we thought, and write letters.

Near the end of that walk, Andy asked a question: does the idea of the common good have meaning in the field of ecology? I didn't have time to answer fully at that moment. And so, the first letter was born. The conversation that had played out after church and in college classrooms moved into an old-fashioned medium:

PREFACE

letters. Letters that we printed out. And mailed. And read. And considered. And responded to.

Those letters became this book—a book of letters between an ecologist and a pastor exploring how ecology and theology, working together, might have wisdom to offer. That conversation is the heart of the letters. But as letters between friends, there's more here than that. The letters are also about life in a time of climate change. About life in a pandemic year. About life.

We are far from the first people to see connections between ecology and theology. Ours are two of many voices exploring how science and religion can work together. And we won't claim to have figured it all out in the course of writing these letters. Our big questions are still big questions. But the bridges that connect this moment and our hoped-for future are taking shape.

And we are grateful to you for joining the conversation.
Andi

Acknowledgments

We thank Middlebury College and its Environmental Studies Program for their generous support of our teaching endeavors, where many of these ideas first surfaced. Andy's sabbatical, which made it possible to bring this book to fruition, was generously supported by the Congregational Church of Middlebury, UCC. Verlyn Klinkenborg, Mary Evelyn Tucker, Bill McKibben, Donald Anderson, and Julie Cadwallader Staub were gracious readers of these letters, when we wondered if anyone else might like to listen in. We are grateful for the keen editorial eyes of Elizabeth Davis, who copyedited the final manuscript. And to our families and friends who have supported and encouraged our book-shaped dreams, we say *thank you*.

JUNE 4, 2020

National Oceanic and Atmospheric Administration (NOAA) scientists reported today that atmospheric CO_2 in May reached 417.1 ppm. This was the highest monthly reading ever recorded at the Mauna Loa observatory.[1]

1. National Oceanic and Atmospheric Association, "Rise of Carbon Dioxide Unabated."

JULY 27, 2020

Dear Andy,

It was lovely to see you last week, and to walk and talk our way along the fields and forests. I wonder sometimes how long it will take, in the post-pandemic future that is surely out there, for the simple fact of physical presence to stop feeling precious beyond all words. I kind of hope it never does; the not-taking-simple-things-for-granted feels like an unexpected gift.

I've continued to think about the question that you left me with at the end of that walk: does the idea of the common good have meaning in the ecological realm? A few days of mulling it over, and the response I gave you in the moment—"yes, unambiguously"—still rings true. Yes, unambiguously . . . and it's complicated. Complicated in a good way—in a way that leads me to think that the richest answers to that question may lie at the borderlands between ecology and theology rather than wholly in either.

My yes is rooted in the core insight that ecological science offers about the nature of our world: that it's a deeply, intrinsically interconnected place. Your question prompted me to look up Ernst Haeckel's original definition of ecology, written in the 1800s. He defined ecology at its birth as "the science of the relations of the organism to the environment."[1] All life is inherently relational—connected. That's both the starting premise of ecology and the product of more than one hundred years of doing ecology.

Interconnectedness. It's one of those insights that's simultaneously obvious and easy to miss, astonishing and easy to take for granted, all at once. The reality to which ecology points us is one in which our very lives depend on other species. Our bodies are built of atoms and molecules that have passed through countless generations of other organisms. The oxygen we breathe is the

---

1. Egerton, "History of Ecological Sciences," 222–44.

July 27, 2020

exhalations of millions and millions of years of growing plants. Ecology asks us to see that all of us, all living things, are woven together by visible and invisible flows of energy, matter, nutrients, even the plant gametes that the checkerspot butterflies outside my window are ferrying from one coneflower to the next. In the moments when I manage to pause and really notice that reality, I feel thoroughly wonderstruck. Too often, though, I don't notice it. Interconnectedness is the fundamental nature of our existence, but it's easy not to see it. I wonder sometimes if it's the very boundedness of our bodies—the embodied illusion that we're somehow discrete entities—that makes it easy for us to believe instead in our own autonomy, our separateness.

As I write this, I'm thinking about the field onto which we were looking as we sat and talked last Wednesday. That field is a pretty simple ecological community, but even so, it contains dozens and dozens of species: plants and animals, fungi and microbes living together, interacting, depending on one another. Even in a relatively simple community like that one, the scope of those interactions borders on bewildering. The daily drama of that field involves eating and being eaten, competition and cooperation, sex and death, disease, symbiosis, decay, and the cycling of atoms around and around and around. I sometimes wish that somebody would invent ecological goggles that would allow us to see those connections, those interactions, more clearly. I imagine that if we could see each of those interactions as a thread tying pairs of interdependent organisms together, the field would start to look like a fabric. And better still, we'd discover, sitting there, that those threads draw us in, too. We're part of that interdependent web of ecological connectedness—all bound up in the midst of it, whether we recognize it or not. As I imagine it, the fabric would not be a smooth, perfect, factory-made bolt of silk or satin. It would be hand-woven homespun, woven by loving hands from threads of myriad thicknesses and textures, representing stronger and weaker connections between pairs of organisms. I'm remembering that beautiful stole that Mary wove for you—I think that's what ecological connectivity would look like if it were a fabric.

July 27, 2020

But to return to your question, I think that the idea of the common good as it exists in the ecological realm flows from that reality of interconnectedness—and the interdependence that arises from it. Try this on for size as an ecological definition of the common good: no species can exist apart from other species, so whether we like it or not, we, the denizens of planet Earth, need one another. We are ultimately all in it together. As I make this claim, I anticipate objections. Surely, I am overstating interconnectedness. Surely, there are species that we could do without. Maybe. But I feel like the sheer complexity of ecology—and the magnitude of what we don't know—argues for humility in the face of that objection. If I can turn this letter over to Aldo Leopold for a moment, he's got some words to say on that score. This is from his essay "Round River."

> The outstanding scientific discovery of the twentieth century is not television, or radio, but rather the complexity of the land organism. Only those who know the most about it can appreciate how little is known about it. The last word in ignorance is the man who says of an animal or plant: 'What good is it?' If the land mechanism as a whole is good, then every part is good, whether we understand it or not. If the biota, in the course of aeons, has built something we like but do not understand then who but a fool would discard seemingly useless parts? To keep every cog and wheel is the first precaution of intelligent tinkering.[2]

Is it just me, or does he remind you just a little bit, in this passage, of God speaking to Job from the whirlwind? So yes, I'm going to stand by my earlier assertion. We are all in it together, interconnected, interdependent.

Things get trickier from there, ecologically speaking, but also really interesting. When I think about the idea of the common good, I think about it as the opposite of self-centered behavior. I can act in ways that center my own interests, or I can act in ways that further the common good. Am I going to go on as many airplane flights as I want, irrespective of the impact on climate change?

---

2. Leopold, *Sand County Almanac*, 190.

July 27, 2020

Or, am I going to choose to forgo my desire to travel for the sake of mitigating climate change? That contrast, between self-centered and other-centered behavior, makes abundant sense in the realm of human decision-making.

It's harder to apply ecologically, though. Is it meaningful to say that a red-tailed hawk, soaring over that field near where we were sitting, is self-centered for seeking to snuff out the life of a meadow vole? It's self-interested, but I don't think I'd be inclined to assign a value judgment to a hawk in the same way that I would to a human who consistently chooses self-interest over the interests of others. In other words, the "good" part of "common good" is not free of ecological complexities. Different species have different "goods," at least in the short term and at the scale of individuals. The red-tailed hawk might well define "good" as having lots of meadow voles to eat today, while the meadow voles would surely define "good" as not getting eaten by a hawk today. Nature's full of conflicts of interest like that, and it's full of self-interest. To complicate it further, what's good for an individual might differ from what's good for a group of individuals: it's good for an individual hawk to eat as many meadow voles as it can, but it's really not good for the population of hawks to run out of meadow voles. The population benefits from a restraint that is not in the individual's interest to exercise. (That reality creates quite an evolutionary puzzle. But that's a tangent for another day.)

Ecological interdependence doesn't have to involve altruism or self-sacrifice (although it can and does). Interdependence is a reality that undergirds a system full of creatures eating other creatures, full of death, full of disease, full of competition and limited nutrients. And yet, we're all in it together. Both things are true. And both things can be true, I think, because there are checks and balances within ecological communities that normally prevent one species from consuming all the resources. That's why invasive species do so much damage: they're often free from the factors—predators, competitors, diseases—that keep them in check in their native habitat. We don't need an individual red-tailed hawk to decide to forgo lunch in order to avoid extinguishing the population of meadow voles. Such conflicts of interest are resolved, for the most part, by the very interconnectivity that I started with. If the red-tailed hawks start to eat too many meadow voles, prey gets scarce, and either the hawks' reproductive success goes down (fewer hawks eating voles) or they switch to some other prey item (fewer voles getting eaten). The hawk doesn't need to *decide* to further the common good.

Humans, however, have a unique (as far as we know) ability to engage in ethical reasoning and make choices that appear to be against our own immediate, short-term desires. We can choose the common good even if it conflicts with our perception of our own self-interest. But here's the thing. I'd argue that, ecologically speaking, the common good—the flourishing of all life on Earth—is absolutely, categorically, unambiguously in our self-interest. If we're all in this together, then we don't thrive if other species don't thrive. So maybe what ecology offers to this conversation is the somewhat paradoxical insight that the route to self-interest lies in caring for species other than ourselves. In other words, the dichotomy between "self-centered" and "other-centered"—between acting for myself or for the common good—doesn't make sense ecologically. If we can't live without other species, acting in ways that allow them to thrive is the very definition of our self-interest.

Maybe the ecological truth is that our self-interest (at least over the long term) and the common good are the same thing. The issue is not that they are in conflict, but rather that we fail to see our own self-interest clearly because we fail to see our interconnectedness. We see instead our short-term desires, and we mistakenly believe that they are the same thing as our self-interest. We fail to see that living as if we were the only species that matters is a dead end, ecologically speaking. But now that I've written that, I wonder if that's really a uniquely ecological insight. How do you understand the distinction between self-interest and the common good *theologically*?

I'm obviously drifting here toward (if not entirely past) the limits of ecology. Maybe the ecological reality of an interconnected and interdependent world shares a border with the theological and ethical concept of the common good. And maybe that borderland—I'm so tempted to call it an ecotone—is a place where

July 27, 2020

we can put our understanding of the world as it is alongside what we know and can dream of the world as it ought to be. Maybe that borderland is the place where we come to understand what it means for life to flourish.

A long walk through such a place sounds like just the thing right about now.

Peace,
Andi

AUGUST 21, 2020

Dear Andi,

Grace and peace.

The letter you wrote reads like the well-spun web of connections it describes. My goodness. Thank you for your tour of ecological kinship. How a question like mine can yield a letter like yours is beyond me, but it's clear I need to ask more questions.

Near the end of your letter, I felt a rush of pride at the recognizable sound of "ecotone." *Wait—ecotone—I know that word!* A few years back, on Easter evening, I tucked that word inside a short poem about Christ appearing and disappearing, like a firefly, where field and forest meet. Where field and forest meet: ecotone. Where two biological communities exchange gifts. Anyway, I like the word. I also like the idea of paddling on the brackish water of Common Good Pond, where ecology and theology can swap stories.

So, on Monday, I went to Kingsland Bay with Ella, Mary, and Rachael. Dad Days in the summer are the best. On the way up Route 7, we stop at the junction store. I go in for provisions. There are a dozen or so people in there. Then, a thirty-something white male walks in past the MASKS REQUIRED sign. He's not wearing a mask. As I see it, everything about his body language is saying, "Nope. Not wearing a mask." The rest of us keep looking at him without trying to be obvious. Meanwhile, he's walking up and down the mini-aisles looking like he wants someone to get in his face about responsibility so that he can get in theirs about freedom. It doesn't happen. He pays and walks outside to his idling truck.

As I replay that scene, your question back to me seems particularly relevant: "How do you understand the distinction between self-interest and the common good *theologically*?" Up front, I want to say this question of yours has made me freshly aware of my need to check in with "self-interest." What *does* that word mean theologically? Does it mean that based on all things I could be interested in, I am most interested in myself? Does it mean selfishness? Self-preservation? Survival? Is it an expression of self-worth? Self-love? Is it why our planetary house is on fire? Is it why we want to put the fire out? Is it possible that "self-interest" is roomy enough to hold all these moods and meanings?

From what I understand about the spread of Covid-19, it would have been in my self-interest to ask the guy at the store to put on a mask. Then again, given how pissed off he appeared to be, it may have been in my self-interest to keep quiet. And while I don't presume to know what the maskless man was thinking, I do wonder: was it in *his* self-interest to be unencumbered by a mask? Did he not care about the other people in the store? Does he think it's a hoax? (Still?) Or did he forget his mask, in haste, on his way to drop in on his girlfriend's mother, who's been feeling lonelier than usual these days?

Self-interest is complicated—as complicated as we all can be. Which is why I had to look up the word when we got home from Lake Champlain.

The main definition tilts toward selfishness: "one's personal interest or advantage, especially when pursued without regard for others."[1] However incomplete this definition may be, the phrase "without regard for others" chafes my understanding of the common good. The common good has to do with conditions necessary for everyone and everything to flourish. And self-interest, by definition, undermines these conditions.

Maybe that's why the word "self-interest" makes me uneasy. I think that this feeling springs from my core belief that "love your neighbor as yourself" *is* the common good—and it's how we get there. And self-interest can get in the way of this.

Self-interest brings to mind something in me that wants to grab first, give second. *Without regard for others?* O Lord, I've been there lots of times. I've been caught and carried by the current of self-preoccupation. I've rejected—wittingly and unwittingly—the truth that I am part of an interconnected, interdependent reality

---

1. Stevenson and Lindberg, "self-interest."

beyond the half-acre of my own little life. I know way more than I'd like to admit about the habits of self-absorption. I've been really far from home, deep in prodigal territory. And despite two dozen years of standing in pulpits and connecting the dots between the common good and the Kingdom of God, what have I really done to be of use to my neighbors whose needs are deep?

These markings of self-interest are mine. But not mine alone. The human tendency to bend toward selfishness and away from the common good has been a thing for a long time.

Maybe that's why Israel's prophets went ballistic over corrupt systems that used poor and powerless people, and why the prophets were constantly telling the high and mighty to get over themselves. And maybe that's why St. Paul wrote to the church in Philippi: "Do nothing from selfish ambition or conceit, but in humility regard others as better than yourselves. Let each of you look not to your own interests, but to the interests of others. Let the same mind be in you that was in Christ Jesus" (Philippians 2:3–5, NRSV).

When I read Scripture, I often recognize the needful reminder to aim my interest beyond the end of my nose. The same goes for reading theology. In fact, I just finished Walter Brueggemann's book, *Journey to the Common Good*. The book begins:

> The great crisis among us is the crisis of "the common good," the sense of community solidarity that binds all in a common destiny—haves and have-nots, the rich and the poor. We face a crisis about the common good because there are powerful forces at work among us to resist the common good, to violate community solidarity, and to deny a common destiny.[2]

Powerful forces at work among us? Indeed. *Within* us, too, as far as I can tell. So, sages of our tradition—Aquinas and Ignatius, Rauschenbusch and Brueggemann, to name a few—keep turning our attention to needs beyond our own. And *The Westminster Catechism* keeps asking, "What is the chief end of man?" And it keeps answering, "Man's chief end is to glorify God, and to enjoy him forever."[3]

---

2. Brueggemann, *Journey to the Common Good*, 1.
3. Westminster Assembly, *Westminster Shorter Catechism*, 1.

Now, when it comes to arresting climate change or living in harmony with the great web of life, I would agree that the idea of self-interest can *include* us acting in ways that allow the species we are dependent on to thrive. But things get tricky when self-interest makes our personal good (survival) the end goal and finish line. Theologically, this doesn't work so well. As I see it, a Christian's pursuit to serve her neighbor and care for creation leads to God—our chief end.

Does this mean self-interest is anathema to one's spiritual life? No, I don't think so. There's a hue of self-interest that matches self-love. And of course, self-love is God-stamped. Love your neighbor *as yourself*. Self-interest is assumed in the Great Commandment. I sometimes forget that.

A few years ago, a sweet man approached me after worship and asked me why I left out "as yourself" from my benediction. I had said, at service's end, something like, "Love God. Love your neighbor. Go in peace." This man shared with me, with equal parts earnestness and gentleness, how much he needed to hear permission to love himself again. Oh, yes. Right. Thank you.

Had I assumed that self-interest (self-love) simply *is* and that the common good (neighbor love) *ought to be*? Maybe so. But that kind soul helped me see that we cannot avert the "crisis of the common good" if we fail to include ourselves as objects worthy of holy love. After all, we are parts of an interconnected, interdependent reality. And for us to be protectors and conveyors of that beautiful truth, we need to include ourselves in the work of the common good—just not at the expense of everyone else at the junction store.

I wish you well, my friend,
Andy

AUGUST 26, 2020

Dear Andy,

Greetings, my friend. I hope that you and your clan are well, and that you've enjoyed a few more Dad Days of summer before we turn that corner into autumn. It got down into the 40s here last night—and there's a box of fall-semester books from the divinity school bookstore arriving for me today. Autumn can't be far behind.

I was happy to get your letter the other day. I was eager to hear your thoughts on the theology of self-interest and the common good. And I was relieved that the US Postal Service is still functioning. Your letter did not disappoint; you chart a lovely path through that thicket of questions about the nature of self-interest.

And it makes my heart glad to hear your affirmation, from a theological vantage point, that the ecological reality of interdependence and the theological idea of the common good really do sing a beautiful harmony together. Your words read like a field guide to the different species of self-interest. That taxonomy adds some important nuance to my ecologist's sense that self-interest is more complicated than mere selfishness. It's also clear that theology can guide us through that taxonomy to a place ecology can't quite reach: past self-interest to the "chief end" toward which our lives are pointed.

That reminds me of how much I've struggled in my life as a scientist with the question of teleology—more specifically, with a mandate communicated to me early on to avoid teleology. I was under review in my second year at Middlebury, and Chris Watters, then the elder statesman in the department, came to visit a class and then invited me to stop by his office so he could share some feedback. I remember that conversation vividly, Chris's face peering out from between the towers of books that grew out of every horizontal surface. I have an equally clear memory of what he told me: I had the makings of a fine teacher, but I needed to reign in my tendency to speak teleologically. Biology is not pointed at a *telos*—a "chief end." The evolutionary process isn't purposeful in that way. I took it to heart, but I've never really stopped wondering whether we sometimes draw that line too emphatically—too universally. Especially within the field of ecology. That remains an open question for me. Anyway, the point is that theology is able to move freely into conversations about meaning, about purpose, about the "chief end" toward which our lives point. And I was grateful for that in your letter because it felt like the completion of an incompleteness in my own thinking.

Our gaze matters. That's what's sticking with me from what you wrote, and I think that leads us somewhere important. Ecology tells us that the common good is in our self-interest, but theology helps us see that self-interest as an end won't get us to the common good. It won't get us there because our gaze matters. Or, as you have said to me more than once, what we give our hearts to matters. We can't get to the common good if we're only looking at the end of our own nose. We can't get to the common good if our hearts love only ourselves. The common good is in our self-interest, but self-interest can't get us to the common good: it's a wonderful little paradox.

And it leaves me with a question: how do we—the collective "we" of humanity, especially those of us who have benefited the most from exploitative living—broaden our field of view sufficiently that self-interest is not "the end goal and finish line," as you put it? How do we do that in a culture in which virtually every message coming our way tells us that furthering our own desires is exactly what the end goal and finish line is? I'm not sure. But the metaphorical half-acre in your letter planted the seed of an idea about what the step toward figuring it out might look like. And it has to do with language—and with metaphor.

On the way to visit my father last weekend, I swung through New Haven to make sure that my sweet little divinity school apartment in Bellamy Hall was in good shape to weather another few months of my absence. I had left some books on my shelf when I departed in March, and a few of them were clamoring for me to take them on a trip to Vermont. Among those was Wendell Berry's

August 26, 2020

book *Life Is a Miracle*. Greeting it like the long-lost friend that it was, I sat right down on the floor and started flipping through it. I found this sentence: "It is impossible to prefigure the salvation of the world in the same language by which the world has been dismembered and defaced."[1] I'm sure I must have emailed you that sentence the first time I came upon it. And probably the second, knowing me. It's one of those sentences that contains more wisdom than I would have believed mere words could carry.

And so, I've been wondering: what language *would* help us to imagine the salvation of the world? It occurs to me that that question is another way of thinking about our work together: the quest for a shared language between ecology and theology is really a quest for a language capacious enough to allow us to imagine the salvation of the world. This thought led me to an article that made the case that metaphors are important tools for scientific discovery.[2]

Metaphors are important tools for scientific discovery.

I love the idea. The authors' point is that metaphor gives us ways of talking about new ideas, ideas that live beyond the limits of our current language. It's one of those surprising sentences that speaks a truth I've long lived with but never noticed. Ecology is a metaphor-rich science; talking about complexity demands metaphor. My last letter to you was full to the brim of webs and nets and fabrics—metaphors that are so commonplace I sometimes forget they are metaphors. What's so cool about framing metaphor as a tool for scientific discovery is that it invites us to look at how the prevailing metaphors change as we discover new things—and to consider which new metaphors might equip us for the journey ahead.

And there's a fascinating history to ecological metaphors. In the early 1900s, an ecologist named John Davidson used the metaphor of art to argue for preservation of wild landscapes. "Just as we preserve the works of great masters, . . . so we are seeking to preserve the works of the greatest of Masters,"[3] he wrote in a report for the Ecological Society of America. A little bit later, Frederic Clements, a

---

1. Berry, *Life Is a Miracle*, 8.
2. Olson et al., "A User's Guide to Metaphors," 605–15.
3. Ecological Society of America, *Preservation of Natural Conditions*, 13.

plant ecologist, talked about an ecological community as an organism. A community, he argued, develops and grows to maturity (the "climax community") just like an individual organism.[4] A community is bound together in the same way that the organs in an individual organism are bound together. It's a lovely metaphor, one that helped us to see the interdependence of the natural world. But it had its limits, and Clements's organismal metaphor eventually gave way to metaphors rooted in engineering.

The opening salvo in that transition came in 1935 in a paper written by an ecologist named Arthur Tansley. The paper, titled "The Use and Abuse of Vegetational Concepts and Terms," was published in *Ecology*, the discipline's flagship journal. It's famous because it is where he introduced the idea of the ecosystem. It was a watershed moment in ecology. But I don't think I had ever read it carefully, front to back, until this week. And the paper not only advances a new metaphor for ecology, it is explicitly about metaphor: its "use and abuse," the title tells us. The paper is a scathing critique of his colleagues, including Frederic Clements. Tansley believed—not wrongly, I think—that ecology was overdue for a new metaphor better able to describe new ideas for how the natural world worked. But he also takes aim at Clements and his friends for confusing metaphor with reality—for doing bad science, in effect. And he makes that point by using another metaphor: religion. Religion is his metaphor for bad science. I had to read it twice to make sure I wasn't just making that up, but he's pretty clear. He writes this about Clements and company: "For them, the plant community . . . is an organism, and he who does not believe it departs from the true faith."[5] At one point, he refers to Clements as a prophet and accuses his supporters of displaying "apostolic fervour."[6] (It's clear he does not intend that to be a compliment.) It's an astonishing paper: a valid claim that we need a new metaphor sufficient for our new ideas paired with a critique of old metaphors that comes in the form of a metaphor equating religion with bad science. I wish I had

---

4. Clements, "Nature and Structure of the Climax," 252–84.
5. Tansley, "The Use and Abuse of Vegetational Concepts and Terms," 290.
6. Tansley, "The Use and Abuse of Vegetational Concepts and Terms," 285.

August 26, 2020

thought of it when we were teaching our J-term class. There's an abundance of fuel here for our discussion of the breakup between science and religion.

Anyway, Tansley ends the paper by proposing a new metaphor, one that he believes captures more fully what we know about how ecological communities behave. The living and nonliving parts of the world are, he says, components of a system. The idea of the ecosystem starts there. I'm so used to the word "ecosystem" as a biological concept that I forget that the core idea of the "system" is rooted in the world of machines and engineering. Tansley's idea that ecological communities can be understood as systems opened the door to decades of discovery of how the living and nonliving parts of the world interact. It laid important groundwork for the modern environmental movement. It also reduced living creatures, air, water, and soil to components of a system: parts of a machine. Things we can control.

Language matters. Metaphors both reflect and shape our thinking: tools for discovery, indeed. And as I reread your letter, the metaphor of your half-acre kept returning to my field of view. It got me thinking about neighborhoods. Obviously, the use of "neighborhood" as a metaphor is not new, but I'm wondering if there's room to push that metaphor more deeply than it's been pushed. It's a metaphor with deep roots theologically—loving our neighbors, as you pointed out in your letter, is one-third of the Great Commandment. (Wendell Berry, in his essay "The Idea of a Local Economy," uses the phrase "the practice of neighborhood"[7] to describe what it means to live in a way that acts out our love of our neighbors. Isn't that lovely?)

But the metaphor of neighborhood also has roots in ecology. The term "ecological neighborhood" refers to the area that matters to a particular organism—the place across which it affects or is affected by the environment, including its fellow creatures. So it occurs to me that if we graft the theologically rooted sense of neighborhood together with its ecologically rooted sense, then maybe we start to have a way of talking about the world that could help us lift

---

7. Berry, "The Idea of a Local Economy," 260.

our gazes a bit and find that path toward the common good. (And having spent too much time the past few days considering the ways in which Tansley's religion-as-bad-science metaphor contributed to that destructive divide between science and religion, the idea of grafting together the wisdom of theology and the insights of ecology does my heart good.)

My ecological neighborhood, in this era of climate change, is arguably the whole world. Every time I burn fossil fuels, I have an impact—small, but not zero—on the entire globe. In fact, studies of vulnerability to climate change have shown that the most vulnerable people—people exposed to climate hazards who lack the resources to respond—are not in the places that have contributed most to the problem.[8] The ecological notion of neighborhood tells me that the world is my neighborhood. And the theological notion of neighborhood tells me that I need to draw that neighborhood close to my heart and care about it. That grafted metaphor of the eco-theological neighborhood simultaneously broadens my gaze beyond my own half-acre out to the rest of the world and brings the rest of the world close to my heart. When I think about it like that, it feels to me like that metaphor might just turn out to be a spacious enough place in which to imagine, as Berry implores us to do, "the salvation of the world."[9]

Peace,
Andi

P.S. That poem you mentioned? The one about Christ, fireflies, and ecotones? If you're ever of a mind to share, please do. Just the idea of it is making me smile.

---

8. Byers et al., "Global Exposure and Vulnerability."
9. Berry, *Life Is a Miracle*, 8.

SEPTEMBER 7, 2020

Dear Andi,

Peace be with you on this gusty Labor Day afternoon.

I'm aware of what day it is, which has not always been the case during the pandemic. Today's the day before the first day of school this year—a later start than usual. Rachael will begin seventh grade tomorrow from Gwen's home office.

This fall has a wildly different look and feel to it, but Labor Day has a familiar ring. When we were kids, Labor Day's main job was to put an end to summer and send us back to school. It marked the start of a nine-and-a-half-month march over blacktop and fields of study. I still think of Labor Day this way. I haven't earned an academic credit since the Mesozoic Era, but the impulse to buy new pens seizes me every September. This is school time.

This is school time for *you*, my friend. I hope you're enjoying the first days of the semester, despite this learning-from-a-distance business. I look forward to hearing about it. Please send dispatches from Verlyn Klinkenborg's virtual classroom when you have the chance.

∞

It's been a full-hearted week, with Ella heading back to Davidson College and Gwen driving Mary to the Boundary Waters for an Outward Bound wilderness adventure. After six months of daily togetherness, I'm trying to cope with not being all together. And I'm not the only one. As I write, Rachael is out on the front steps strumming chords and singing "Crowded Table" by The Highwomen. Little sister's coping.

∞

September 7, 2020

Your last letter arrived about the time I started tilling Sunday's sermon. It took me a while to come to terms with the fact that I was writing a sermon, not a letter back to you. And there's a good reason for this. Paul's call for neighborliness in yesterday's Epistle lesson (Romans 13:8–14) finds a deep response in your letter. I loitered in Romans 12 and 13 last week. I read your letter. I loitered some more. I reread your letter. There's plenty of "neighborhood" in those hills, where your letter and Paul's meet.

I'm thinking about the trove of ecological metaphors you dug up. Thank you for that! The metaphors are rich and seem to say a lot about their authors. Ecological community is sacred art to be preserved. Love that. Ecological community is an organism inherently bound together. Intrigued. Ecological community is a machine-like system, an "ecosystem." Suspicious. Then, your metaphor of an ecological neighborhood. Yes, please. I like that one. It sends me back into Paul's letter.

In chapter 12 of Romans, Paul lays out the principles and practices of "neighborhood." *Love one another with mutual affection,* he says (v. 10). Also, *contribute to the needs of the saints* (a.k.a. the poor), *and extend hospitality to strangers* (v. 13). Also, *rejoice with those who rejoice and weep with those who weep. Live in harmony with one another,* says Paul (vv. 15–16a). *So far as it depends on you live peaceably with all* (v. 18). Why? Because that's what "love your neighbor as yourself" means.

Then, in Romans 13, Paul approaches the summit of neighbor-love on a different trail. He says that "love your neighbor as yourself" is not only about doing good for someone else. It's also about not hurting others. Paul makes clear what the Ten Commandments make clear: loving our neighbor as we love ourselves includes some "don'ts" and "shall nots." As my beloved teacher/mentor/friend David Bartlett wrote, "love does not mean doing whatever feels good to me. It means—at the very least—living up to the law, which will not do the neighbor harm. Love may be more than that, but it is never less than that."[1] (I miss David more than I can say.) David was dialed into Paul. Paul was dialed into Moses. The second tablet

---

1. Bartlett, *Westminster Bible Commentaries: Romans*, 119.

of the Law is all about how *not* to live in the neighborhood. Don't kill, don't cheat, don't steal, don't lie, don't covet. Why? Because that's no way to "love your neighbor as yourself."

All in all, Paul is devoted to the flourishing of communities where people care for each other and find courage in each other's company to tell their lesser angels to hush. Paul's "neighborhood" is a place of deep communion. In his letter to the Galatians, the Apostle says that Jews and Greeks, slaves and free, men and women, are *one* in Christ (Gal 3:28). Not 1A and 1B. Just one. So, who's left out? Nobody. The Pauline neighborhood is as large as it is revolutionary. It is global in scope. By the grace of God, *all* are worthy of welcome. Paul's sense of community includes everyone because (as Jesus makes clear) our neighbor is anyone.

∞

So, yes—as a reader of Paul and follower of Jesus, I can say without qualification that the map of the world is my neighborhood, too. And the neighborhood is burning.

There are a whole lot of people living in harm's way today, due to rising seas, dying fisheries, killer heat waves and droughts, devastating fires and floods, and a bullpen of other disasters. And there are a whole lot of people among them who are really, really short on options.

So, how can I purport to love them as myself if I don't set out to do them some good and *not* do them harm? What good *can* I do? What can I do to not make things any harder? These are questions worth asking . . . and answering. What to do and not to do?

∞

The wind is carrying Rachael's voice past my open window. I hear her missing her people and hitting the notes of a love that abides. We sow and grow love. We garner and give love. "We" is key at the round, crowded table, where each one brings something to the neighborhood potluck and everyone's a little happier to live here.

Happy Labor Day,

On Easter Evening

Andy

P.S. Here's the "ecotone poem" you asked for.

ON EASTER EVENING

Fireflies hang
On the edge
Of the ecotone.
Light comes and
Goes near
The deep blue
Woods, like
Someone I know.

SEPTEMBER 13, 2020

Dear Andy,

Happy Welcome Sunday to you! I was thinking about last year's Welcome Sunday while driving to Hinesburg this morning. My heart ached more than a little bit thinking about the simple loveliness of that day: the whole beloved church, sitting at tables, sprawled on the grass, eating and passing the time. But then I thought about you and the kids and an ice cream truck gathering behind the church this afternoon. And that made me glad.

I hope the Nagy-Bensons are adjusting to being a trio. We're on the cusp of going down to a duo again. So, yeah. The image of Rachael sitting on the front steps singing "Crowded Table" is echoing around my heart. I'm right there with her. Galen just left for his dad's house; from there, he's moving into my friend Heidi's house—she's returning to British Columbia until the spring semester, at least, and wants a caretaker. He's going to buy their old VW, too. Perfect set-up for a twenty-one-year-old. He's been busy reassuring me that his delight at the prospect of moving out and having his own transportation isn't personal. I've been assuring him that launching his own life is exactly what he should be doing—exactly what I want him to be doing. And I'll miss him. Both are true.

So is this: I was really happy to get your letter. It's one to savor. My ears perked up, especially, with your description of the capaciousness of the Pauline neighborhood. The whole world matters: that's music to this ecologist's ears. We don't get to pick and choose what we care about. Yes, yes, yes. I love that we arrived at that same point along different paths.

And like you, I'm deeply worried about this neighborhood of ours. The West Coast is on fire. The Arctic's thawing—and on fire. The oceans are rising. There are more hurricanes this year than there are names. And those are just this week's headlines. I find myself reading them with disbelief, even though I can't tell you how many grant proposals and papers I wrote over the years that started with a litany just like that. "Here are the things that are likely to happen if current trends continue," I'd write.

God, have mercy.

Now that it's happening, I have the unsettling feeling that I never fully processed what I was saying in the dry language of academic science. Knowing what I know—knowing what I have known for years and years—I shouldn't feel surprised at what I'm reading in the news. But I do. Day in and day out, I lived in the shadow of a future that is now upon us. So why am I surprised that it's here? I think I coped by keeping it all at arm's length. I wrote that litany, over and over again. I understood exactly what it meant. But I couldn't let myself feel the grief it contained.

I feel it now.

Of course I feel it. I became a scientist because I loved this world. I don't remember a time in my life when I wasn't head-over-heels in love with this world. And the world with which I am in love is suffering. Grief is to be expected. (Somehow the grief, too, catches me by surprise.)

So, your question is also mine. What are we to do? My efforts at an answer start with a story. I had a colleague named Matthew Sturm—he's a geophysicist who studies snow. He's also a brilliant storyteller. He and I were in a meeting together, along with a bunch of other scientists who studied arctic climate change. We were trying articulate in plain language what we collectively knew to be true about climate change in the Arctic. Somewhere along the way, though, we got stuck in the thicket of what we *didn't* know. To get us unstuck, Matthew told us a very simple story. It went something like this:

Imagine the earth system as a boulder resting on the top of a mountain. We know that we're pushing on that boulder. And we know that it's starting to roll. We don't know exactly where it will stop. That's true. But you're forgetting—he told us—*that we know one more thing beyond a shadow of a doubt: we know that the boulder is going to end up somewhere different than where it started.*

September 13, 2020

That was enough to get us unstuck. We didn't know the future with perfect clarity, but we did know that it would be different. We knew that the boulder was rolling. We knew that we needed people to pay attention to that fact.

I'm thinking of that boulder as I think about the question of what we should do.

The answer is everything. We should do everything we can possibly do, and probably some things that we think we can't. We need to throw every damn thing we've got in the path of that boulder. Every. Damn. Thing. Because here's the other truth illuminated by Matthew's story of the boulder, the one that ultimately became the story of the paper that came out of that meeting. A boulder rolling downhill doesn't just change position—it accelerates. Things are going to change faster and faster and faster. The longer we wait, the farther that boulder will be from where it started and the harder it will be to stop.

The *why* of that is a consequence of the interdependence of this neighborhood of ours. That interdependence means that ecological systems are full of feedback loops: a change in one thing causes a change in another which causes a change in the original thing. Negative feedbacks stabilize a system by counteracting the initial change. Our bodies are regulated by negative feedback loops. Body temperature rises, we sweat, evaporative cooling happens, body temperature falls back to normal, and stability is maintained. Positive feedbacks amplify the original change. They're destabilizing. Not surprisingly, they're uncommon in physiology. (Childbirth is apparently one of the few examples of a positive feedback loop in human physiology.)

Both kinds of feedback loops occur in ecosystems. The crucial question is: which kind will be more powerful as climate changes? Will negative feedback loops prevail, in which case Earth will stabilize itself—at least in part—and help save us from ourselves? Or will positive feedback loops prevail, in which case the response of ecosystems to anthropogenic changes will accelerate the rate of change? Nobody knows for sure.

At the moment, the positive feedback loops seem to be winning. The Arctic Ocean is becoming ice-free, which exposes dark ocean water, which absorbs more sunlight, which warms the ocean, which in turn warms the rest of the Arctic. Positive feedback loop. The frozen Arctic soil warms and thaws, and long-frozen organic matter starts to decompose, releasing carbon dioxide. Positive feedback loop. The Arctic burns and long-buried carbon is released into the atmosphere. Positive feedback loop. Fire frequency increases and more forests burn every year, releasing carbon dioxide. Positive feedback loop. The cold Arctic tundra becomes hospitable to trees, which move north, and the white winter landscape of the Arctic is suddenly covered with dark, coniferous trees that protrude above the snow and absorb sunlight and warm the surface. Positive feedback loop. I could go on, but you get the point.

There *are* negative feedback loops. In some places, warming temperatures will make plants grow faster. That means more carbon dioxide taken from the air and turned into wood and biomass. On the scale of the whole earth—especially if we include the oceans—that could make a measurable difference. Plants may yet help to heal the damage done. They may, but evidence that they *will* is shaky at best. Right now, the positive feedback loops are prevailing.

The boulder's accelerating.

Before you start to regret opening this letter (am I too late?), there's one more piece of the puzzle. There's another negative feedback loop. It's us. Or, it could be us. Unlike the Arctic Ocean or permafrost, we get to choose how we want to be. So, here's what it could look like. The world starts to burn, thaw, flood, and blow away. If we're paying attention, we'll notice that. And here's where your letter is calling to me loudly. We aren't just thinking creatures, we're feeling creatures. We're a hot mess, as a species, but we've got big hearts. We're made to love each other—made for those crowded tables. And so, it's possible—isn't it?—that when we notice what's happening to the world, we will feel loss. We will remember that we actually love this neighborhood of ours. In the name of love, we will change our ways. (I'm pretty sure you preached that early in the pandemic, in fact.)

A love-motored negative feedback loop: for love, we change our ways. And the boulder slows.

September 13, 2020

It's a good story. I don't think it's impossible. I mean, think about it: we know how to do everything we need to do. It's not like, say, a pandemic caused by a virus for which there is no vaccine, yet. We have the technological know-how to live differently. We just need the will. That's a big "just." I get that. But I think that's where that neighborhood metaphor offers us something powerful.

Your question—"what to do and what not to do?"—is pretty straightforward, ecologically. Ecology is really clear on the fact that limits exist. Whether we want them to or not, they do. And whether we want to or not, we live within them. So, we do live within our means, and we do not live beyond our means and into the means of other people and species. We reject the fantasy that we can have whatever we want. We—especially the "we" that has monopolized more than its fair share of resources—give stuff up. We consume less. A lot less. We travel less. We eat more plants. We farm differently. We teach our children differently. And we change systems. We vote for leaders who will enact the systemic change we need to make it possible for us to live lightly enough on the earth; the Green New Deal is a great start.

Live within our means.

It's easy, conceptually. But where I get stuck as an ecologist is in the awareness that it's a really hard sell. I'm no marketing expert, but I'm pretty sure that "just say no to the luxuries you've come to love" is not a message likely to rally people to the cause. But here's where the round table is so important! Your letter, which I've read and reread almost a dozen times over the last twenty-four hours, is a road map forward. I can tell this story as an ecological story about limits and "no" and giving stuff up. And it is all of those things, and we need to say those things out loud. The story you're telling in your letter also includes those limits (which is really cool). But the story you're telling is not mostly about "don't" and "shall not." It passes through there, but it doesn't stop there. Your story is about love.

If the past six months have taught us anything, they've taught us that in the name of love we can do things we didn't think possible. We can give up things we didn't think we could live without. Sure, there are some holdouts; not everyone is sold on the whole idea of loving our neighbors. But most people are. So that list, a couple of paragraphs up, of what we have to do to save this world? Don't you think it's possible that people would do that for love? I think we know in our heart of hearts that the way we're living isn't working for us. I think our big-hearted selves yearn for something different. I really do.

I don't quite know what comes next. I feel like I'm seeing this path one flagstone at a time. One letter at a time, maybe. But here we are at this place that feels so full of possibility. Ecology and theology are both holding up big neon signs: "neighborhood, this way." And if we're all willing to walk there together, I think that we can heal this place we love. We can heal each other. I don't know how we will get there, exactly, but I know it's where I want to go.

This all has me thinking about your ecotone poem. I'm thinking about the miracle that the light comes. And even more than that, I'm thinking about the miracle that when the light goes and darkness falls, we are capable of imagining the light. We're capable of holding fast to the hope that it will come again.

And I believe it will.

Peace, my friend,
Andi

SEPTEMBER 21, 2020

Dear Andy,

So, dear friend, after putting that last letter in the mail, I found myself lingering on the "God, have mercy" at the top of the second page. I also found myself needing to write my weekly two-page essay for Verlyn K.'s Writing the World class.

And the week's headlines and that prayer combined themselves into an essay. It occurred to me that it's really a postscript to that last letter, so I'm sending it along to you—to be reunited with the letter from which it sprang.

Peace,
Andi

September 21, 2020

More than five million acres have burned in the western United States. The disintegration of two monumental glaciers in Antarctica is accelerating. There were five named hurricanes in the Atlantic on one day. Hurricane forecasters have announced that we've reached the end of this year's list of storm names. The Arctic is thawing. And it's on fire. The Greenland ice sheet may have melted past the point of no return. That's what September has brought. So far. I'm surprised at my own surprise. I studied climate change for twenty-five years. September's headlines are the future I used to write about. "If current trends continue," I would write.

Current trends continued. And now I'm inserting prayers into the headlines as I read. Five million acres burned: God, have mercy. Thawing poles: God, have mercy. More hurricanes than names: God, have mercy. The headlines become a litany of lament. And I wonder at my grief. Not at the why: I wonder why *now*. Or maybe I wonder why not then. Every summer for a quarter-century, I traveled to the edges of the boreal forest—austere places I came to love. I kneeled before spruce trees. I measured them. I thanked them. I counted their rings and learned their histories. I returned home and wrote dispassionate prose about their possible demise.

I didn't grieve then. Learning to study the world, I also learned to love it from a distance. Distance is the language of academic science. When I was first learning that language, I was told that a first-person pronoun was a controversial choice. We were not supposed to be in our science. Also, distance was a way of coping. I recognized that coping in myself only after hearing someone else articulate it. My brother, a conservation biologist, visited my class one day to talk about Bicknell's Thrush. He laid out in detail the many threats it faces. And he laid out in detail the efforts underway to save it. He explained that its future was grim anyway. A student raised her hand at the end. How did he keep from being crushed by the grief of working on a species that might be doomed? "I just try not to think about it," he said. "That's probably not a satisfying answer. But it's true. I don't let myself think about how things are likely to be thirty years from now. I couldn't keep doing this work if I did." I was surprised at my own surprise then, too. It had never occurred to me to ask her question—of my brother, or of myself.

My answer was the same as his. I wrote papers about a grim future. I tried not to think about it. More than that, I tried not to feel it.

Lately I've been asking myself: what if I hadn't tried so hard? The sound of that "what if" keeps me awake at night. What if I'd allowed myself to grieve? What if I had remembered sooner that we are feeling creatures before we are thinking creatures? In ancient Israel, there were mourning women—professional keeners—who were responsible for public lamentation when tragedy happened or disaster struck. What if I had taken up the practice of public lamentation? What if all of us documenting the unfolding tragedy of climate change had taken up the practice of public lamentation? Scientists have started talking about grief—their own and others'. They've named the griefs particular to this moment: anticipatory grief, ecological grief, climate grief, solastalgia. The words sound like a dirge.

I lie in bed at night and half-dream of scientists as keeners. We stand on the streets outside our laboratories. We weep for the losses we've borne witness to in the pages of academic journals and lab notebooks. And we weep for the losses yet to come if humanity doesn't change its ways. Grief rolls off our tongues in a low, keening cry: anticipatory grief, ecological grief, climate grief, solastalgia. The lament begins.

SEPTEMBER 23, 2020

Dear Andi,

Ruth Bader Ginsburg (1933–2020). The wrecking ball of 2020 keeps swinging, doesn't it? And yet, according to a Jewish teaching, someone who dies on or just before Rosh Hashanah is a *tzaddik*, a person of great righteousness.[1] Justice Ginsburg died as the sun was setting last Friday evening. Ruth the Tzaddik worked nobly for all persons under the law and, along the way, taught many men what "all" means. She was a reckoner—she weighed truths. Also, she loved the opera. May her memory be a blessing.

Ruth Bader Ginsburg's passing has me thinking about the space between how things are and how things ought to be, what Parker J. Palmer calls the "tragic gap."[2] As I hold this tension and read your letter and its postscript, I really, really want to speed ahead toward the hoped-for beloved community, to the neighborhood where life-giving "oughts" are realized.

But I can't do that yet.

First, I need to heed one of the lessons I've learned in church. Right relationships with God, neighbor, and all creation require the courage to name what *isn't* right. And what I need to say, whether I want to or not, is that we are slowly, and ever less slowly, making it impossible for living things to live here. That isn't right. That's sin. The Intergovernmental Panel on Climate Change (IPCC) says with near certainty that anthropogenic climate change is the result of human activity.[3] This sounds to me like an occasion to pray. "Most merciful God, we confess that we have sinned against you in thought, word, and deed, by what we have done, and by what we have left undone."

1. Harris, "Does RBG's Rosh Hashanah Death Really Make Her a 'Tzadik'?."
2. Palmer, "Standing in the Tragic Gap."
3. Intergovernmental Panel on Climate Change, *Climate Change 2014*, 5.

September 23, 2020

In the Gospel of Matthew, the first followers of Christ are called to love God and to love their neighbors as themselves, and to do to others as they would have others do to them. But they (like us) miss the mark.

The number of yearly climate-related disasters has tripled since the 1990s.[4] Last month's litany of lament echoed across the land—California wildfires, the derecho that did damage in South Dakota and Ohio, the Atlantic hurricanes. Lives and livelihoods were lost at summer's end because the planet is reeling. Same thing happened last year. And the year before that. And the year before that. "Positive Feedback Loops Prevailing on Earth": if this were a headline, it'd sound better than it should to a distracted ear. The truth it tells, like the story you tell, is dire. Climate change is a rogue boulder accelerating downhill, threatening humankind. And most of us helped push it over the crest of the knoll. God, have mercy, indeed.

So, your letter and postscript make me wonder why. Why aren't human beings doing everything we can to slow the boulder? Why aren't I? Why have the "current trends" continued, decade after decade? On restless nights lately, I've been drifting down and up these paths looking for clues:

> Maybe some of us don't know how *not right* things are.

> Maybe some of us do know, at least enough to know better, but we can't deal—or we're dealing with other stuff, like teaching our kid to drive and figuring out what to make for dinner for the zillionth time.

> Maybe some of us have a favorite uncle who calls bullshit on global warming every time it snows.

> Maybe some of us believe that God will pluck us out of the burning house.

> Maybe some of us believe that we will be saved by technological advances.

---

4. Oxfam International, "5 Natural Disasters That Beg for Climate Action."

September 23, 2020

Maybe some of us are so in love with this world that we have to look away because the heart can only take so much ache.

Maybe some of us are scared witless and don't know what to do.

Maybe some of us are profiting like crazy and pouring gas on the fire and planning to retire early beside a rising ocean.

Maybe some of us don't want to give up what makes us happy or our place at the center of the universe.

Maybe some of us—especially the billion of us scraping by on a dollar a day—are just trying to survive.

Maybe some of us don't want anyone telling us what to do.

Maybe some of us are waiting for someone to lead us and tell us what to do.

Maybe some of us would care more if we knew our neighbors better. It's hard to love what you don't know.

In any case, after another fitful sleep and Green Mountain sunrise, climate change is still real. It's time to do everything we can to slow that boulder down. The "why" of that is very clear: if we do not change our daily ways, and if the oil and gas sector keeps reeling in profits of $11 billion a year (2019),[5] and if we keep electing people who have an apparent allergy to caring for the well-being of our planet, human civilization will not survive the positive feedback loops of our making.

I think that's why I'm praying more these days. Not because I think prayer alone will stop climate change, and not because I think God will do the work that is ours to do. (God does not do the dishes that we leave in the sink.) I think I'm praying more because

---

5. Meredith, "Oil Major Totals Full-Year Profit Falls."

September 23, 2020

that is something I *can* do. Mostly, I pray because I need God's help, and I want to be helpful for God's sake. I pray because that's when, if I can stop all the chatter in me, I hear the faint assurance of God-with-us and remember God's delight in the diversity and breathtaking beauty of this world. I pray to speak my heart and to listen for wisdom. I pray to be honest to God. Like, I'm scared to live in Trump's hellscape. Like, I'm heartsick for my kids' future on this planet, let alone the future of their kids. I pray because I know my need for mercy and grace, and I trust that God wants to give us a garland instead of ashes.

Speaking of prayer, do you remember sharing Walter Rauschenbusch's gem (from 1910 or so) with our class last year? The prayer called "For This World"? It bears repeating:

> O God, enlarge within us the sense of fellowship with all the living things, our little brothers, to whom thou hast given this earth as their home in common with us. We remember with shame that in the past we have exercised the high dominion of man with ruthless cruelty, so that the voice of the Earth, which should have gone up to thee in song, has been a groan of travail. May we realize that they live, not for us alone, but for themselves and for thee, and that they love the sweetness of life even as we, and serve thee in their place better than we in ours.
>
> When our use of this world is over and we make room for others, may we not leave anything ravished by our greed or spoiled by our ignorance, but may we hand on our common heritage fairer and sweeter through our use of it, undiminished in fertility and joy, that so our bodies may return in peace to the great mother who nourished them and our spirits may round the circle of a perfect life in thee.[6]

This prayer reminds me: I still want to reach that beloved community, that neighborhood where life-giving "oughts" are realized.

---

6. Rauschenbusch, *Prayers of the Social Awakening*, 47.

SEPTEMBER 23, 2020

And like you, I still believe that if we're all willing, we can heal this place . . . with God's help.

Blessings on you, my friend.

Peace,
Andy

SEPTEMBER 30, 2020

Dear Andy,

The morning that your most recent letter was delivered into my hands, I spent some time sitting on one of the exposed boulders at the top of the field behind my house. I was practicing looking at the leaves. When the first rays of sunlight hit the edge of the field, it looks like the trees have swallowed the sun. It's ridiculously beautiful. But lately, I find myself looking *through* the beautiful colors. All I can see is worry: worry that the early arrival of foliage season is yet another sign of this epic drought, worry that it's warm enough in late September for me to be barefoot. If I'm not paying attention, I can go through my day like that—harvesting worry from beauty.

So, I was trying, sitting there on that rock, to see the world in all of its complexity: to see the terrible things, yes, but not only the terrible things. Beauty matters, too. Surely, it does. And I don't need to tell you that that project is not easy these days, as—thank you for the phrase—the wrecking ball that is 2020 swings.

And into that train of thought flew your letter. I hear you wrestling with the same thing: how to hold the beautiful along with the terrible, the truth of grace with the reality of sin, hope with despair. I'm grateful for that sentence of hope at the end. It felt hard-won—a hope that has outlasted illusions. And I'm glad, too, for the reminder that pervasive though the impact of sin may be, our story did not begin with sin and our story surely won't end there. God wants for us a garland, not ashes. I needed that reminder.

∞

My friend Dawn Jefferson preached her senior student sermon in Marquand Chapel—well, the Zoom version of Marquand—on Monday. She's a brilliant preacher. She preached on Moses in the wake of Miriam's death. Her point was one I've heard you make, that grief will find its way to the surface, one way or the other. Better, therefore, to acknowledge it openly. So, in the middle of the sermon, she asked people to type into the chat the name of someone they were grieving—someone lost during this pandemic. The names scrolled by too quickly to see. So many names, so much grief. Folks stayed together on Zoom for a full half-hour after worship just to be together. In prayer, in grief, in community. It was a holy time—and it was healing. Far more healing than I could have imagined of anything happening on Zoom.

In the middle of her sermon, Dawn said a sentence that I've been carrying with me: "We are called to look ahead, past this wilderness, to a Joshua generation."[1] It feels like another way of articulating what I was trying to get at above. We have to see this wilderness time for what it is—to use the language of your letter, we have to name what's not right. But if we are called to see past the wilderness of this moment, and I think we are, that can't be all that we see. We won't get out of the wilderness if wilderness is all we allow ourselves to see. That truth is captured by your list of answers to the question of why we're letting that rogue boulder that is climate change accelerate. Some of us, your list tells me, need stories to get us going—stories that we can believe in, stories that allow us to look at the hard truth, stories that speak into being the rugged, unquenchable hope that we carry. The Israelites had God's promise that there was an end to the wilderness. And they had stories about what that end looked like. It looked like Canaan, the land flowing with milk and honey.

What stories could *we* tell about what lies beyond *our* wilderness? What does hope look like? Walter Rauschenbusch's prayer in your letter is pointing there. And I think that's why it matters to me to not forget to see beauty. Beauty is a reminder of what that future toward which we're walking might look like. Beauty lifts my gaze and makes it possible to maybe, just maybe, see beyond this wilderness.

∞

1. Jefferson, "When Miriam Is Missing."

September 30, 2020

You know that book I emailed you about last week? *The Hebrew Bible and Environmental Ethics*? Mari Joerstad, the author, argues that mourning and drought are linked in the prophetic texts. Drought is how the land mourns, she says. Where I might throw myself on the ground and weep, the land withers and dries up. Listen to this:

> That the earth suffers because of human sin is the dark side of the much celebrated interconnection and interdependence of humans and the world in the Hebrew Bible. ... The earth's mourning in the Prophets gives voice to the suffering of nonanimal nature, and chastises humans for putting other creatures through pain, adversity, and death.[2]

I don't have trouble believing that the earth is telling us right now that it's suffering. That it's mourning. Joerstad makes the point that in the Hebrew Bible, non-human creation—including the parts we'd consider inanimate—exists on its own terms. It has a relationship with God apart from ours. And it's affected by what we do, not just in strictly material terms, but in a way that is responsive to violations of the harmony and integrity that God intends. When we miss the mark, the land mourns our sin. It mourns the breach to ecological integrity. I think we need to hear that word, "mourn," with deep humility—to not read ourselves and our own understanding of mourning into it too much. But *that* the land mourns? I can imagine that. And I'm captivated by the idea.

I don't think her point is inconsistent with what ecological science tells us. The non-human world has a life apart from us. It's communicative. It's responsive. It's relational. There's nothing that I know as an ecologist that would argue against those things. In fact, it seems to me that ecology is a way of listening to what the earth is telling us. And what ecology tells me right now sounds an awful lot like the earth crying out because of our sin.

This raises a question for me. What else is the earth telling us? Are there other stories? Ones that might help us to see what lies

---

2. Joerstad, *The Hebrew Bible and Environmental Ethics*, 144–45.

September 30, 2020

beyond this wilderness? If I turn down the volume of the worried chatter in my head, I do hear another story.

∞

Up Chandler Hill Road from my house, there's a stone marker that marks the birthplace of Fanny Collar. The story, as I've heard it, goes like this. More than a decade after the town of Ripton was chartered, it remained unoccupied. The town's founders were worried about their profits. The rumor spread that a free plot of land would be given to the first white family to have a child in Ripton. Fanny Collar's father apparently thought this sounded like a great opportunity. Sometime in the fall of 1801, he convinced his very pregnant wife to head up to Ripton. (Can you imagine how that dinner-time conversation went?)

Fanny's mother apparently agreed, and Fanny was born in November, right up the road from me. "First white child born near this spot in Ripton." There are volumes of history—mostly tragic—folded into that sentence. It marked a watershed moment in Ripton's history. Ecologically, things changed quickly after that point. An 1871 atlas shows Ripton with five sawmills, a shingle and clapboard mill, three schools, a church, a parsonage, and a grist mill.[3] It was practically a metropolis compared to today.

Between those sawmills and the sheep pastures that are attested to by the network of stone walls that still lace the woods around here, it's a safe bet that those were not good years for the forests of Ripton. The state as a whole went from mostly forested to mostly deforested over that time period. It's hard to imagine now: Ripton, deforested. And then the sheep industry collapsed. Sawmills were abandoned. People moved out. If I were parachuted into that scene—forests gone, soil eroding into streams—I would have been hard-pressed to tell a hopeful story about the ecological future of this place. I would have seen a story of ecological grief: a story of the demise of old forests, the likes of which we'll never see again.

And then something amazing happened. The forests grew back. It's easy to take that for granted. I went for a walk in the woods

---

3. Beers, *Atlas of Addison Co. Vermont*.

after writing that sentence. I noticed the forests, everywhere. I noticed the stone walls, slowly collapsing back into the forest floor. I noticed how tall the beech trees were above them. And I just let myself feel joy. The forests grew back. Somewhere—I've lost track of the source—I came upon a reference to a journal written by a woman who'd grown up in the nineteenth century. She was writing toward the end of her life, in the early 1900s. She describes traveling from Ripton into Middlebury and seeing the view slowly obscured by regrowing trees. And she describes her grief at that change. Ecological grief. She'd grown up in a world of pastures. She grieved the loss of those open vistas—of the world as she'd known it.

Where she saw heartbreak, you and I might see resilience, and that's one of the stories this land has to tell. It sits alongside the story about grief. They're both true. Yes, the forests are different now, to be sure. The trees are smaller. There are more of the early-successional species than there used to be. But that's changing. The white ash in the forests behind our house are dying, giving way to sugar maples and beech—shade-tolerant, late-successional species. Next time you walk up the Skylight Pond trail, keep an eye out for red spruce in the understory. They're late successional species, too, slowly returning to the places they used to grow.

This land has suffered. And this land has healed.

Although Ripton's forests tell a story of hope, the resilience of the past is not an ironclad promise for the future. Climate change still threatens to upend everything. But if, as the prophets seem to suggest, humans and the land can mourn together, isn't it possible that we can hope together, too? And wouldn't it be good to have company in the hoping?

I don't know exactly what lies beyond this wilderness time. I can't quite see that far down the road yet. But I find myself wanting to tell the story of what *could* lie beyond the wilderness, a bit further down the road—the story of what could be if we start righting what's wrong. I think we need stories to move us forward. And when I think about telling the story of what lies beyond this wilderness, I think it might begin here. With forests laced with stone walls. With a neighborhood—a beautiful neighborhood of human and non-human creatures and the land on which they live—grieving and

September 30, 2020

hoping and healing together. I'm not sure where it goes from there, but that feels like a place to start.

And wherever this road leads, my friend, I'm grateful for your company on it.

Grace and Peace,
Andi

OCTOBER 20, 2020

Dear Andi,

The work-life balancing act has been leaning hard toward work this month. Thanks for your patience with this slow-train-coming reply. Thanks, too, for your last letter. Your invitation to swap stories of hope was much needed. Many thanks, friend.

Anyway, after I preached last Sunday's sermon, I realized that my message might also be a response to your letter. So, here's the sermon, "Walking With Faith," based on 1 Samuel 17:32–49.

∞

A friend of mine has been making deliberate choices lately to actively listen for new and old stories of hope. She talks about it in terms of protecting her heart and soul from getting stuck under the weight of all that threatens to undo us. There's nothing really prescriptive in what she says. But I find good medicine in her descriptions of hope. So, I've taken up listening more closely, too. And I've been looking, daily, for glimpses of a world that is healing and healthier and actually whole. Here's a glimpse:

Earlier this month, I spied an online headline in *The Guardian*, a British daily newspaper—"Cambridge University to divest from fossil fuels by 2030."[1] In the article, the university's chief investment officer says: "Climate change, ecological destruction and biodiversity loss present an urgent existential threat, with severe risks to humankind and all other life on Earth. The investment office has responded to those threats by pursuing a strategy that aims to support and encourage the global transition to a carbon neutral economy." It is fair to say that the investment office did not come up with this on its own. For the past five years, a student-led campaign

---

1. Taylor, "Cambridge University to Divest from Fossil Fuels by 2030."

called Zero Carbon has been pushing and pressing school officials to remove its oily investments from their $9 billion endowment.[2]

The decision to do so was complicated by the fact that Cambridge has held close financial and research ties with major fossil fuel companies. In fact, twenty years ago, British Petroleum donated twenty million pounds to found the BP Institute at the University. Then, just last year, Cambridge accepted a six-million-pound donation from Royal Dutch Shell.[3] But with pressure from Zero Carbon, the university made the important decision to divest from Big Oil and to build up significant investments in renewable energy.[4]

That's a newsworthy decision. And really—it came about relatively quickly. I mean—yes—five years is a lot of lost time in the midst of a crisis. But let's put this divestment move in context. Cambridge University was founded in The Year of our Lord 1209.[5] But women were not admitted there until 1882. And women were not awarded degrees by the university until 1948.[6] So, it took five years for a student-led campaign to convince the university to divest from the fossil fuel industry. It took five years for the students, and some professors, too, to sway the university to establish a goal of net neutrality before the next decade dawns. Five years. That's 148 times faster than the university's divestment from male privilege. One hundred forty-eight times faster. That's like driving 4,440 miles per hour in a 30 mph zone.

I share this story because climate change does present an urgent existential threat, and because if we don't change our ways, we stand little chance of surviving the feedback loops of our making. But most of all, I share this story because we need old and new stories of hope to find our way.

---

2. Taylor, "Cambridge University Urged Again to End Fossil Fuel Investments."

3. Gayle, "Cambridge Accepts £6m Shell Donation for Oil Extraction Research."

4. Common Dreams, "Cambridge University Makes Historic Break With the Fossil Fuel Industry."

5. Cambridge University, "Early Records."

6. Chambers, "At Last, a Degree of Honour for 900 Cambridge Women."

October 20, 2020

This morning's Scripture lesson is on that list. For all its testosterone and violence, the story of David and Goliath is a hope story at its core. In the epic theater production of David's dynamic life, today's passage might be considered Act One, Scene Two. In the opening scene, right before today's reading, David is chosen by the prophet Samuel to be Israel's future king. But here in chapter 17, that future's a ways off. David's just a boy, and Israel is in trouble now.

They are presently at war with the Philistines. And one of the Philistine soldiers, named Goliath, has the soldiers of Israel quaking in fear, wringing their hands, sowing doubts among their ranks. Simply put, Goliath is a problem. A big problem. A well-armed and armored problem. An intimidating, trash-talking problem. Now, David—the shepherd boy—gets a good look at this mountain of a man and no doubt learns a few choice words. Then, David volunteers. "I'll fight him," he says. His older brothers laugh, like older brothers do. But David isn't kidding.

King Saul steps in. He says to David, "You are not able to go against this Philistine; for you are just a boy." But David presses and pushes and changes the king's mind. So, Saul gives his own armor to the boy. And David gives that a try, but it's much too big, way too heavy. He can barely walk, let alone fight. David takes off the armor. Then, he takes off toward Goliath with not much more than a shepherd's sling and five smooth stones.

David doesn't ask for a crash course on sword fighting before he heads down the hill. He brings what he has, and he does what he knows to do. He may be young, but he's been keeping his father's sheep safe from predators for a while now. He's been training for this moment without knowing it. And well, you know the rest. The fight is over in a flash—like those hyped-up bouts that end with a flurry of punches in the first round.

David slings a stone. Goliath falls. The Philistines flee. Goliath goes down in defeat. But he's been coming around ever since, in one form or another. Every one of us has looked up at a problem that was bigger than us. Each of us has gulped. Each of us has said "No way" when there looked to be no way.

October 20, 2020

But listen carefully, and you might hear young David saying something faithful.

What I hear David saying is this: God is bigger than our biggest problem. And God equips us with our own set of tools, gifts, and strengths to overcome the challenges we face. What I hear David saying is true.

Walking with faith is about remembering not to mistake our oceanic God for a pond. Walking with faith is about using every bit of what God gives us to take on the giants that threaten to undo us. That list can seem daunting some days. But God is still God. And we are, even now, holding in our hands five smooth stones.

That sounds to me like a host of hope stories waiting to be told.

Amen.

Peace,
Andy

OCTOBER 28, 2020

Dear Andy,

I was glad to get my hands on your "Walking with Faith" sermon. There's so much to love there. David and Goliath? That's a story for this moment. I love the way you excavate down to the hope-filled center of the story, for one thing. It's a good reminder that hope doesn't always come in quite the package we'd expect. Of all the hopeful things in your sermon, there's one that has held onto me. It's this sentence: "And really—it came about relatively quickly." "It" is divestment. Relatively quickly is the lightning-fast 148-fold acceleration of institutional change.

I suppose it struck me because so much of the Goliath-like character of climate change is about its speed. Fast, and accelerating. But with that sentence, you reminded me: the boulder that is climate change isn't the only thing that can accelerate. Forces of good can, too. I needed to hear that. That sentence put me on the lookout for other examples of hope-as-acceleration.

And I think I found one. While cooking dinner the other night, I was half-listening to an interview with an ecologist named Terry Chapin. He was talking about his new book—*Grassroots Stewardship: Sustainability Within Our Reach*. I started paying attention when I heard him say: "I'm certain that we can shape the future." That sure sounded like hope to me.

But before I go there, I need to tell you a little bit about Terry Chapin. He was an important mentor to me early in my career. He taught me a lot about how to be a good scientist and even more about how to be a good human. But the best thing to tell you about him is that when Galen was little, he thought Terry was the Lorax. Not *like* the Lorax—he thought he was the actual Lorax. For good reason. Terry looks more than a little bit like the Lorax. And Galen usually encountered Terry when we were out in the field in Alaska.

October 28, 2020

He associated Terry with trees. One day, Terry met us at a field site up in the Brooks Range and hid behind a tree to surprise Galen. He popped out with a big smile on his face and Galen shrieked with delight—in that awesome little-kid way—and shouted: "it's the Lorax!" The big smile on Terry's face would be where he and the Lorax part ways—Terry is impassioned, but he leads with kindness and joy instead of cantankerousness. He's good people.

Terry studied climate change in the Arctic starting way back in the 1980s, long before today's consensus on the issue had emerged. It was Terry who pulled me into the world of Arctic climate change when I finished my PhD. At his urging, I dove into deeper scientific waters than I would have on my own. He saw something in me that I didn't see in myself. Encouraged me when I needed an encouraging word. He's a special human being, and when he says things, I pay attention. As one does when the Lorax speaks.

So, back to that future he thinks we can shape. A while back, Terry got really interested in the bigger picture—in how science could make a difference in the way people interact with non-human nature. I've lost track of his work in the last few years, so I got curious about what he's been up to. That interview that I half-heard led me to a talk he gave in Sweden last fall[1]—just after winning the 2019 Volvo Environment Prize. He started with hard reality. If we fail to act, he said, the earth will become "significantly less habitable than today." He talked about his own grandkids when he said those words. He made it clear, with a vulnerability I don't hear in scientific talks, that climate change is personal for him. He went on to describe the world that he believes we can still create. A place of cooperation, not competition. A world where we care for each other and everything else. A world where the word "wealth" doesn't just mean money—it includes natural, social, and human capital. A "stewardship economy," he calls it. It was lovely.

Then he started talking about how to get there. He acknowledged the need for systemic change, but his focus these days is on individual behavior. I appreciated hearing his reasons for that focus. You and I have grappled with the worry that individual changes

---

1. Chapin, "Triggering Transformation to Sustainability through Stewardship."

just aren't enough. That's what the pull of despair feels like for me: the feeling that anything I can do is too little. Terry's a wise person, and my hunch is that he's making a deliberate effort to combat exactly that despair. But it's not just that. His effort is grounded in science that tells him that individual choices do matter. He made that clear. Individual choices—transformative actions within our own little lives—can push the needle. Especially if they happen within the lives of those of us in developed countries. Take one less trans-Atlantic flight. Eat a plant-based diet. Drive less or not at all. He framed this in terms of happiness—our need to "pursue non-consumptive dimensions of happiness." Go outside and celebrate nature—fall in love with the world.

Those choices are all familiar. What struck me about his talk—what's still striking me—is how he explained *why* he's talking about individual behavior. I heard three reasons. First, Terry believes in the good in people. He really does believe that we're capable of doing the right thing if given the chance. There's a beautiful hopefulness to how he sees humanity. It's healing, in these days of division, to listen to someone who looks out at the world and sees the good. I hadn't realized until I was sitting there listening to him with tears in my eyes how much I needed to hear that. The second reason is that he thinks individual humans are capable of more wisdom and more long-term thinking than governments and businesses. Our families point us toward an inter-generational timeline. We can think about our children's future and our grandchildren's future and be motivated to act on their behalf.

The last reason he shared connects me back to your sermon. When you and I talk about individual choices as a component of the solution, I think we both get stuck on the issue of scale. We are so very small (David!) relative to the size of the problem (Goliath!). Terry went down that road. Yes, individually we are small. But he added, we're not isolated. We're not alone, marooned on tiny islands of autonomous thinking. We're social creatures, connected into social networks. I've been struggling with how to connect the fact of our connectedness to the pragmatic reality of climate change. Terry's talk closed that loop for me. Whether we admit it or not, he said, we're affected by what our neighbors do. Even when we don't

OCTOBER 28, 2020

love them very well, we still want to be like them. Social connectedness—and specifically our desire for social conformity—can be the agent of change that allows individual actions and decisions to be globally transformative. I admit, I adore the idea of conformity driving change. Transformative actions in an individual's life can spread rapidly along social networks. (Not in the sense of social media, but actual flesh-and-blood social networks.) I think what I heard him saying is that not everybody needs to be convinced by scientific data or moral arguments that climate change warrants a vigorous response. That conviction can spread because people listen to and want to emulate their neighbors—it can spread because of a desire to conform.

I got excited. And so, instead of doing the reading I was supposed to be doing for class today, I dug into that idea. And I came upon a really intriguing paper that came out last month in the journal *Nature Climate Change*.[2] It's about how households in Papua New Guinea make environmental decisions in the face of climate change. What makes a household take "transformative action"—substantive changes in how they live that make them more resilient to climate change? Neighbors. People are embedded in social networks and those networks can amplify change. So, an external actor—an NGO, a government—can seed the idea that change is needed. That change then spreads through a social network, even among people who ignored or didn't encounter that "external actor."

They found that people who took transformative action were also affected by their connections to nature. Folks who live and work in close proximity to nature—those who have "social-ecological ties"—were more likely to take transformative action to address climate change. And it turns out that the combination of those two things—high social connectivity and high connectivity to ecological networks—amplifies change. Folks were more likely to transform their lives when their neighbors did so and when they or someone they knew was strongly connected to non-human nature. Those effects were far stronger than whatever those external actors were doing.

2. Barnes et al., "Social Determinants of Adaptive and Transformative Responses to Climate Change," 823–28.

October 28, 2020

I read all that while thinking about that boulder rolling toward us. Thinking about the forces that are needed to counteract it and—back to your sermon—about where there might be accelerating forces of good. Terry's talk and this paper are pointing at one: our connectedness creates the potential for change to accelerate. Change will start slowly in such a system, but then it will take off.

And so, I started looking at the divestment question that you raised. I think that's what we see there. Maybe this is cynical of me, but I would hazard a guess that Cambridge University administrators ultimately acted not because they suddenly discovered that divestment was morally right. I assume they acted because that movement—an expression of social connectedness—created social pressure to which they felt they had to respond. That works both within and among institutions. The more institutions divest, the more social pressure there is to conform and divest. In the first ten years of the twenty-first century, exactly one institution of higher education divested from fossil fuels in the United States. In the second ten years, twenty-six did so. I read in a report by 350.org[3] that there was a 22,000 percent increase in assets committed to divestment—across all sectors of the economy—between 2014 and today. Across a total of 1,245 institutions, $11 trillion have been removed from fossil fuels. I'm no economist, but that's a lot of dollars. The forces of human good—powered by social connectedness that organizes itself into social movements—are accelerating. Terry mentioned in his talk that there are, at this moment, more than two million organizations and groups worldwide working on issues of social and environmental justice. Savor that number for a minute. Two million. That's the largest social movement in history. A lot of connectedness. A lot of potential for amplifying the good.

Could it be that our connectedness is the stone in our pouch? Could we be on the cusp of living out a version of the David and Goliath story in which—surprise!—there are millions, maybe billions, of Davids, all intent on slaying Goliath? That story is deeply hopeful and points at the work that needs to be done. How do we strengthen social networks? How can we build the strong social-ecological ties

---

3. 350.org, "350 Campaign Update: Divestment."

October 28, 2020

that were associated with transformative climate action in Papua New Guinea?

Terry ended his talk by saying that we "inhabit a planet in transition. Its trajectory is ours to choose."

I spy hope, my friend. The pull of fear and despair is so strong right now. And yet. These letters are my reminder that hope is stronger still. Hope is always stronger still. It's out there, shining brightly—it hasn't been hard for either of us to find it. Which is something. Anyway, I'm grateful for these letters—and more grateful than words allow for your company along this road.

Peace,
Andi

NOVEMBER 18, 2020

Dear Andi,

Thanks for your letter. I've read it a few times now. Who doesn't need refills of hope these days? With the pandemic rearing up and the nation breaking up, hope has been laboring up a steep pitch lately. Mention of green shoots during this stick season could be met with forgivable suspicion. And yet, you pull it off beautifully.

I especially enjoyed your introduction of Galen's Lorax. The words you use to describe him—your mentor—include good, kind, joyful, thoughtful, hopeful, impassioned. We all could use a Terry Chapin in our lives.

Thanks, too, for following Terry's lead and tracking down the paper in *Nature Climate Change*. I didn't expect to go to Papua New Guinea when I opened your envelope, but I'm glad I did. What I brought back with me was the encouragement to look for hope in two million small groups working overtime to beat climate change. Small groups of individuals making changes for the good of the whole community. Sounds like church to me.

I read a piece the other day about a small group of leaders from the United Church of Christ who played a formative role in launching the environmental justice movement. In the late 1970s and early eighties, a fierce cohort of UCC folks directed the gaze of church and state toward the intersection of environmental degradation and racism. Faithful witnesses like Dollie Burwell, Rev. Leon White, and Rev. Dr. Benjamin Chavis, Jr., along with others from the UCC's Commission for Racial Justice, got into some "good trouble." Working closely with the mostly poor, mostly Black residents of Warren County, North Carolina, these church leaders shined a light on the dumping of newly banned toxic chemicals (polychlorinated biphenyls) in that area. Chavis called it what it was—environmental racism.

November 18, 2020

In September '82, this small group led a six-week protest laced with acts of civil disobedience. Dozens of arrests the first day—over five hundred before it was over. This became a national news story.[1] And it worked. The PCB-laden site was detoxified. Eventually, laws were written and enacted, including President Clinton's Executive Order called "Federal Actions to Address Environmental Justice in Minority Populations and Low-Income Populations."

Small groups of individuals making changes for the good of the whole community. Yes, please.

So, there's more. The public witness in Warren County was only the start of our denomination's coordinated effort to frame environmental woes in terms of the inequities of race in America. In the late eighties, Chavis and the UCC's Commission for Racial Justice published a landmark report called *Toxic Wastes and Race in the United States*.[2] The report tells a hard truth: the racial composition of a community was a key variable in determining the placement of toxic waste facilities across the country. At that time, three out of five Black and Hispanic Americans lived in communities that the EPA called "uncontrolled toxic waste sites."[3] Again, environmental racism. (By the way, Justice and Witness Ministries continued the work of the Commission for Racial Justice by publishing *Toxic Wastes and Race at Twenty: 1987-2007*.[4] Unsurprisingly, the racism flushed from the bushes in '87 was still nesting in poor communities of color twenty years later.)

Then in 1991, our denomination's Commission for Racial Justice sponsored an event in Washington, DC—it was called the First National People of Color Environmental Leadership Summit. More than six hundred people attended the four-day event. At the gathering, seventeen principles of the environmental justice movement were adopted. This is where they landed:

> WE, THE PEOPLE OF COLOR, gathered together at this multinational People of Color Environmental

1. United Church of Christ, "A Movement Is Born."
2. Commission for Racial Justice, *Toxic Wastes and Race*.
3. Commission for Racial Justice, *Toxic Wastes and Race*, ix.
4. Bullard et al., "Toxic Wastes and Race at Twenty," 371–411.

November 18, 2020

Leadership Summit, to begin to build a national and international movement of all peoples of color to fight the destruction and taking of our lands and communities, do hereby re-establish our spiritual interdependence to the sacredness of our Mother Earth; to respect and celebrate each of our cultures, languages and beliefs about the natural world and our roles in healing ourselves; to ensure environmental justice; to promote economic alternatives which would contribute to the development of environmentally safe livelihoods; and to secure our political, economic and cultural liberation that has been denied for over 500 years of colonization and oppression, resulting in the poisoning of our communities and land and the genocide of our peoples, do affirm and adopt these Principles of Environmental Justice:

1. Environmental Justice affirms the sacredness of Mother Earth, ecological unity and the interdependence of all species, and the right to be free from ecological destruction.

2. Environmental Justice demands that public policy be based on mutual respect and justice for all peoples, free from any form of discrimination or bias.

3. Environmental Justice mandates the right to ethical, balanced and responsible uses of land and renewable resources in the interest of a sustainable planet for humans and other living things.

4. Environmental Justice calls for universal protection from nuclear testing and the extraction, production and disposal of toxic/hazardous wastes and poisons that threaten the fundamental right to clean air, land, water, and food.

5. Environmental Justice affirms the fundamental right to political, economic, cultural and environmental self-determination of all peoples.

6. Environmental Justice demands the cessation of the production of all toxins, hazardous wastes, and radioactive materials, and that all past and current producers be held strictly accountable to the people for detoxification and the containment at the point of production.

7. Environmental Justice demands the right to participate as equal partners at every level of decision-making including needs assessment, planning, implementation, enforcement and evaluation.

8. Environmental Justice affirms the right of all workers to a safe and healthy work environment, without being forced to choose between an unsafe livelihood and unemployment. It also affirms the right of those who work at home to be free from environmental hazards.

9. Environmental Justice protects the right of victims of environmental injustice to receive full compensation and reparations for damages as well as quality health care.

10. Environmental Justice considers governmental acts of environmental injustice a violation of international law, the Universal Declaration on Human Rights, and the United Nations Convention on Genocide.

11. Environmental Justice must recognize a special legal and natural relationship of Native Peoples to the U.S. government through treaties, agreements, compacts, and covenants affirming sovereignty and self-determination.

12. Environmental Justice affirms the need for urban and rural ecological policies to clean up and rebuild our cities and rural areas in balance with nature, honoring the cultural integrity of all our communities, and providing fair access for all to the full range of resources.

13. Environmental Justice calls for the strict enforcement of principles of informed consent, and a halt to the testing of experimental reproductive and medical procedures and vaccinations on people of color.

14. Environmental Justice opposes the destructive operations of multi-national corporations.

15. Environmental Justice opposes military occupation, repression and exploitation of lands, peoples and cultures, and other life forms.

16. Environmental Justice calls for the education of present and future generations which emphasizes social and environmental issues, based on our experience and an appreciation of our diverse cultural perspectives.

17. Environmental Justice requires that we, as individuals, make personal and consumer choices to consume as little of Mother Earth's resources and to produce as little waste as possible; and make the conscious decision to challenge and reprioritize our lifestyles to ensure the health of the natural world for present and future generations.[5]

Two decades later, the relevance of this group's work is both a problem and an opportunity, isn't it?

Anyway, I'm with you and Terry and the Papua New Guineans. And I'm compelled by the idea that small groups of neighbors are change agents. I also love that small groups from within our small denomination have been doing this work for a while. I spy hope in this. Our church's commitment to care for God's creation and to love our neighbor as ourselves started long before either of us could spell "polychlorinated biphenyls." It's helpful to me to remember that. The roots of hope run deep, and they help me stand up and speak up for environmental justice now.

Small groups making changes for the good of all.

∞

A couple Saturdays ago, I drove into Burlington after dropping off Rachael at her acting class. The cars were thick. I rolled down my windows—an unseasonable 73 degrees and sunny. Up ahead,

5. Lee, *Proceedings: The First National People of Color Environmental Leadership Summit.*

November 18, 2020

I heard honking horns. Lots of them. More car horns than I've heard in total in Vermont. The city sounded like a colossal flock of Canada geese. I saw bodies hanging out of car windows and holding up signs and shouting with joy and recording the moment on their phones.

As I rolled down College Street, I realized I was in a spontaneous victory parade. People were four, five deep on the sidewalk cheering the honking cars. UVM students were well represented. Thumping tunes were coming from Church Street. People were dancing. I tried out my horn and waved and passed through the beautifully wild party—and just like that, something in me started to hurt less.

We elected Joe Biden to be our next president. We elected Kamala Harris to be our next vice president. This was the news. We picked the ones who have sense enough to bring "hope" and "science" together in their victory speeches. And we picked *against* something, too. We are not as bad as I feared. That's the thought I kept thinking. We are not as bad as I feared. It's not the most hopeful thought I've ever had, but honestly, I have feared in recent years that we are not as good as I had hoped. And those fears still come to visit.

Driving home that day, I heard on the radio that 70 percent of Republicans do not think the election was fair. Isn't that about the same number of Republicans who do not consider climate change a major threat? I'll have to check out how many registered Republicans there are, but that's got to be something like 35 percent of our population.

Besides thinking that we are not as bad as I feared, I was also wondering if this is what it feels like to get out of an abusive relationship. The bully-in-chief deputized the bullies, and there were enough of them to make it a too-close race. More than seventy million citizens of the United States of America just voted for the incumbent who is endorsed by hate groups all across the land, the guy who still has not conceded the election, the guy who now plans to sell Arctic oil leases.

Hope is a lifelong workout, and I know it has helped me over the years to have a crew with whom to exercise. The power of small groups to make changes for the good of the community, nation, and planet is not to be underestimated. And maybe that's some of what I saw at the intersection of College and Church earlier this month: a diverse group of human beings standing up and speaking up for our future. The collective exhale, the relief, the jubilation—oh, my. Mine was an ugly cry behind a pair of sunglasses.

Can each of us and small groups of us bend the arc of climate change away from disaster? Of course we can. So, I hope.

Wishing you and your family a lovely Thanksgiving.

Peace,
Andy

DECEMBER 26, 2020

Dear Andy,

I started writing this letter exactly four weeks ago. The Saturday after Thanksgiving, to be exact. I had the optimistic idea that I'd manage to respond to you before the confluence of Advent and end-of-semester work hit. Unwarranted optimism, obviously. December, and its steady stream of papers and exams and Advent reflections and sermons and prayers and—oh yeah, life—needed all of my attention. I know you know the feeling.

Anyway, here we are on the far side of a Christmas like no other. I hope you and yours managed some Christmas joy amidst all of the missing of togetherness and crowded sanctuaries and people. Your Christmas Eve sermon was good medicine, and not just because it sent me back to watch *A Charlie Brown Christmas* for the first time in many years. Although, that helped, too. "It's Christmas, Love." Yes, indeed.

The ache of missing the people I love got to me yesterday around midday, so I went for a long, long walk in the pouring rain to shake off the self-pity threatening to set in. I got out to a beaver pond, way out in the back stretches of the Rikert ski trails. There I was: feeling the heartache of this pandemic year, soaking wet, standing ankle deep in mud on snowless ski trails, peering through the cloudiness at the gray Green Mountains. And you know what welled up in my heart? Joy. Surprised by joy. C.S. Lewis got that right. I couldn't help myself. I started singing. "Go up to the mountain, joyful bearer of good news. Shout with a full voice, our God is near!"[1] The monks of Weston Priory would've been proud of the sentiment—if not the tunefulness of my singing. And it's true. Our God *is* near. In the mud, the rain, the ache. Our God is near.

---

1. Monks of the Weston Priory, "Go Up to the Mountain."

December 26, 2020

And I woke up this morning feeling the exhaustion ebb a little bit. Feeling mental clarity start to return. And so, here I go, into the letter I was writing way back when. . . .

∞

I was glad to learn in your last letter about the birth of the environmental justice movement. I didn't know that story. It's beautiful, tangible proof that small groups of dedicated people can make a difference. And it reminds me to broaden my temporal horizon—to look beyond the last few years, or even the last few decades. We are moving forward. Maybe slower than we'd like, but forward all the same. Thank you for the reminder not to put all of my stress on the last few wild years. There's a longer trajectory at work here.

A few days after receiving your letter, I caught the tail end of an NPR story that seemed to travel through a similar hope-filled terrain.[2] It, too, reminded me to take the long view. That things change. Slower than I'd like, but they do change. Anyway, this story involved soil carbon. Soil carbon, and an unlikely alliance between former opponents. (It occurred to me that the very fact that people were talking about soil carbon on the national news is a mark of progress in this fight.) The story they were telling goes like this:

About ten years ago, the American Farm Bureau Federation was active in opposing action to combat climate change. They fought vigorously against a cap-and-trade bill in Congress. They're a major lobbying group—and they give a lot of money, mostly to Republicans. They have a reputation for being more interested in supporting big agriculture than family farms. But back in November, their president announced a new alliance with, among other partners, the Environmental Defense Fund. The alliance is called the Food and Agriculture Climate Alliance, and they're pushing support of federal climate legislation—including the Growing Climate Solutions Act,[3] which was introduced by a bipartisan group. Two Republicans, two Democrats. (I can't help but wonder how

---

2. Charles, "Farmers Are Warming Up To The Fight Against Climate Change."

3. Braun, "Growing Climate Solutions Act of 2021."

different our national discourse might be if we heard more stories about cooperation amidst divisiveness.)

The Alliance advocates for programs to make it more feasible for farmers to switch to renewable energy and adopt practices that increase soil carbon sequestration. Soils are crucial carbon sinks—they store a lot of carbon. The amount of carbon in soils, worldwide, is estimated to be 2.3 times greater than the total carbon in the atmosphere and 3.5 times greater than the carbon in all living plants. A recent study by folks at Stanford found that 70 percent of all of the carbon in soils is in soil affected by agriculture.[4] So, farming matters. Tilling disturbs soils; it accelerates the rate of decomposition and causes soils to release carbon (as CO_2). But what farming practices do, farming practices can also undo. Low-till practices, increased perennial crops—these things increase soil carbon storage. And it turns out, they bring other benefits, too. It's a bit of a win-win situation. So that's what this group is trying to achieve: create incentives (mostly monetary) for farmers to farm in a way that builds soil carbon back up.

On the one hand, it would be easy to point out the imperfections here. Plenty of people have done so—the NPR coverage launched promptly into an interview with someone eager to point out all that was suboptimal about this proposal. The word "voluntary" shows up a lot, for one thing. And carbon offsets, the mechanism for funding some of this, are problematic if they're not done right because the rich get to buy their way out of changing the way they do things. I understand that critique. The stress on monetary incentives makes me uneasy. Monetary incentives got us into this mess! Are we really supposed to believe they'll get us out? I can muster plenty of skepticism, too.

On the other hand, here's what I heard most loudly, especially on the heels of reading your last letter. The long view—the one that measures not only how far we are from where we want to be, but how far we've come—looks like this. Ten years ago, the American Farm Bureau Federation sent farmers to Washington, DC wearing "don't cap our future" hats to protest cap-and-trade legislation. Now

---

4. Jordan, "Soil Holds Potential to Slow Global Warming."

they've teamed up with an environmental NGO to promote climate legislation. Imperfect climate legislation, yes. But still: climate legislation. People who were yelling at each other a decade ago have rolled up their sleeves to try to find a path forward. Doesn't this sound a little bit like small groups of people working together toward a better future?

And in its very imperfection, it makes me think about the panda's thumb. You're gonna have to bear with me while I work my way through a somewhat circuitous thought. I hope it's worth the trip...

Back in 1978, Stephen Jay Gould wrote about the panda's thumb for his regular column in *Natural History* magazine. He used the panda's thumb as an example of what he calls evolution's "Tinkertoy approach."[5] The panda's thumb is not a thumb so much as an extension of an arm bone—"familiar bits of anatomy remodeled for a new function,"[6] Gould writes. It's a sixth digit, their actual thumb having been repurposed at some point into something less thumb-like and more finger-like. So evolution used an arm bone to make a new thumb. Gould offers this as an example of the ways in which evolution is different from good design. Evolution makes use of the raw materials at hand, which may or may not be the ideal raw materials you'd want if you were *designing* a thumb. Evolution's limited that way. It can only work with the materials available to it. And so all of us living things are a little bit jury-rigged. Like pandas with arm bones for thumbs.

The jury-rigged nature of life is one of the things I loved most about teaching evolution. It makes me deeply happy to know that all of life on the earth is infused with this wild, wacky, imperfect, pragmatic, "can-do" spirit that allows arm bones to become thumbs. It reminds me that nature finds a way. It's not always going to be pretty, but nature finds a way. And our stories of hope make me think that our path forward—humanity's, that is—is going to be the same way. The messy mosaic of hope, I think you called it a few weeks back. Yes. Jury-rigged. Not optimal. Probably not pretty—at

---

5. Gould, "This View of Life: the Panda's Peculiar Thumb," 20.
6. Gould, "This View of Life: the Panda's Peculiar Thumb," 28.

least in any conventional sense. Requiring us to befriend our former enemies. Requiring us to keep our eyes on the ideal—and also to be open to the imperfect stepping-stone solutions that we meet along the way.

You know what that kind of hope is? It's *respair*.

"Respair" *is* a word, despite what spellcheck thinks. I heard it on the radio a few weeks back on an NPR show about words.[7] That led me on a trip into the *OED*. As a noun, it means "fresh hope, or recovery from despair."[8] As a verb, it means "to have hope."[9] Both noun and verb forms date back to the sixteenth century. The *OED* traces the word back to a variant of the verb "respire." (To breathe. To have new hope. How great is that?) Turns out that to respire is also—way down the list of alternate definitions—to "have relief from some unpleasant or undesirable situation; to recover hope, courage, or strength."[10] That sense of respire became, apparently, its own word. *Respair*.

I love that respair reminds me that hope knows the bleak and desperate feeling of despair. I love that it reminds me that hoping is like breathing. I wonder if respair is what those jubilant folks you encountered on Church Street were feeling the day the election was finally called. Tired, giddy, joyful. Still worried that the tantrum-throwing occupant of the White House might not leave, but crawling out of despair into new hope.

And the word respair makes me think that maybe hope is like the panda's thumb, too. Maybe hope—hope as it is made manifest in our human hands, at least—is like evolution, in that sense. A little bit jury-rigged. Sometimes inelegant. Limited to making use of the materials at hand, but finding a way forward even where there seems to be no way. Arm bones become thumbs. And a world is healed, piece by broken piece.

The stories of hope that we keep unearthing insist that the brokenness of this world is being met, all over, by efforts to repair.

7. Barnette, Martha and Grant Barrett, "Episode 1557"
8. "Respair, n."
9. "Respair, v."
10. "Respire."

December 26, 2020

To heal. And that's the part that feels so full of grace. As our eyes open—as we see what we've wrought on our planetary home—we feel the shiver of fear, yes. These human hearts of ours ache, yes. They keep us awake at night fearing for our children and our children's children, yes. But then? Then we roll up our sleeves. And we band together. Small groups of dedicated people, some of whom used to stand on opposite sides of the problem, and we fix what we can fix. Which is mostly local stuff. A toxic waste dump in Warren County, North Carolina. Soils in the Midwest. It's like we're made for hope. And right about now? That feels like a miracle to me.

And sure, each of those local stories of somebody fixing something is, on its own, insufficient. But when we zoom out, as I think you and I are doing in trying to piece together these stories, that messy mosaic of hope appears. The cracked and broken world being slowly glued back together again—in jury-rigged fashion, like the panda's thumb.

It makes me wonder if what we're talking about in these letters is a world of righted relationships. Relationships with each other, with creation, with God. I stumbled back into Joseph Sittler's *Essays on Nature and Grace* a couple of weeks ago while writing a paper for my Natural Theology course. He writes that because people "exist and are as relational entities, only a redemption among can be a real redemption."[11] A redemption *among*. The righting of relationships *among*. Do you see that, too, in this mosaic we're making?

Blessings on you, friend, as you walk the path toward Epiphany—and to the rest and restoration waiting for you on the far side of it.

Peace,
Andi

P.S. After I signed off yesterday, I read that the writer Barry Lopez died on Saturday. Such a loss. His book *Arctic Dreams* accompanied me to Alaska. His assertion that we need both science and poetry to genuinely understand the world found me as I realized that "poet"

---

11. Sittler, "Nature and Grace in Romans 8," 178.

was not the path for me—but wrestled with whether science was. His refusal to fall into either/or thinking helped.

There's a passage in it—one that I marked a long, long time ago—that's in helpful conversation with Sittler's idea of "redemption among":

> For a relationship with landscape to be lasting, it must be reciprocal. At the level at which the land supplies our food, this is not difficult to comprehend, and the mutuality is often recalled in a grace at meals. At the level at which landscape seems beautiful or frightening to us and leaves us affected, or at the level at which it furnishes us with the metaphors and symbols with which we pry into mystery, the nature of reciprocity is harder to define. In approaching the land with an attitude of obligation, willing to observe courtesies difficult to articulate—perhaps only a gesture of the hands—one establishes a regard from which dignity can emerge. From that dignified relationship with the land, it is possible to imagine an extension of dignified relationships throughout one's life. Each relationship is formed of the same integrity, which initially makes the mind say: the things in the land fit together perfectly, even though they are always changing. I wish the order of my life to be arranged in the same way I find the light, the slight movement of the wind, the voice of a bird, the heading of a seed pod I see before me. This impeccable and indisputable integrity I want in myself.[12]

Reading that passage was the first time I was able to ascribe words to the longing that lived in the deepest parts of my own heart. The longing that I ultimately realized was about God. Those words are an early stepping stone on that path. Even now, I can't read them without my eyes filling with tears, remembering what it felt like to acknowledge that longing.

When I came upon it while paging through my favorite parts of the book the other night, it occurred to me that integrity is an important word for this conversation. *Biological integrity* is actually a thing that scientists measure—it's a composite index of different

---

12. Lopez, *Arctic Dreams*, 404–5.

December 26, 2020

aspects of an ecosystem that's meant to tell us something about how much of a dent human activity has made in the way things are supposed to be. And I love that it has a dual meaning. The sense of wholeness, and the sense of ethical soundness: we need both.

It seems to me that this messy mosaic of hope that we're trying to envision is all about rebuilding integrity. It's about restoring wholeness to that which is broken. Redemption among. Right relationship. Neighborhood. Integrity. A lexicon of hope, perhaps.

JANUARY 14, 2021

Dear Andi,

I'm not quite sure where to begin.

Happy New Year? Well, yes, I do hope this year holds many blessings for you and your family. Still, those well-meaning words—Happy New Year—sound terribly ironic these days.

Was it only a dozen days ago that the president of the United States, in a recorded phone call with Georgia's Republican secretary of state, pressured that state's top election official to "find" 11,780 votes for Trump's reelection bid?

Was it just last week that the president incited a deadly insurrection of the government he supposedly leads?

Was it yesterday—really?—that Donald Trump earned the ignoble distinction of being the only US president to be impeached twice?

The calendar says so. But honestly, the first two weeks of January have been quite a year.

∞

Last summer, President Trump ensured that Black Lives Matter protesters would be met with unreasonable force.

Last week, President Trump made other arrangements. And lo, the white nationalists' message was clear: Black lives will never matter as much as ours. An Auschwitz hoodie? Who makes that? Who wears that? A Confederate flag inside the nation's Capitol? The world of Trump is, in turns, bonkers and bigoted and built on greed and fooled by the showman.

And now, Senator Lindsey Graham, a white, powerful man, says the last thing we need right now is an impeachment trial. I disagree. As a white, much less powerful man, I think an impeachment trial is the first thing we need right now. It's how we teach our

January 14, 2021

children that words and actions matter, and it's how we continue the necessary work of dismantling white supremacy.

Do I expect that seventeen Republican senators will be able to find their moral compasses in time to convict a crook? No, I do not. I expect the vast majority of Republican senators will talk about healing without accountability and unity without racial justice. They'd have us run a fool's errand. They'd have us pass on the hard work of repairing the breach.

∞

The trouble, of course, is that I need to love them.
All of them.
The duped "stop the steal" mob. The anti-maskers. The Christian nationalists. The white nationalists. The Republican members of Congress and those who elected them. The climate change deniers.

I know I need to love them, even if I don't want to, even if I'm not sure exactly how.

∞

So, if the path forward requires us to befriend our enemies, as you rightly say, then I'll need an extra serving of grace before heading off into those thickets. Grace enough to be generously kind. Grace enough to not mock those who are making a mockery of common sense and electoral math and scientific facts. I will need God's help to see God's likeness in the ones I don't like very much.

> O to grace how great a debtor
> Daily I'm constrained to be!
> Let Thy goodness, like a fetter,
> Bind my wandering heart to Thee.[1]

∞

It seems important to reflect on what "love your enemy" does *not* mean. As far as I can tell, "love your enemy" does not require forfeiting the ideal of the common good. It does not require us to

---

1. Robinson, "Come, Thou Fount of Every Blessing."

January 14, 2021

be ambivalent about the difference between a cup of lies and a cup of truth. It does not mean for us to be silent before injustice. It does not ask us to let slide the violence that rends the garment of community. Far from it.

The other day a friend and colleague, Elissa Johnk, posted an article on Facebook that helped me. (Elissa is the pastor of First Congregational Church, Burlington. If you don't know her, I'd love to introduce you. She's the real deal.) It was an article published in November, a piece by Rebecca Solnit called "On Not Meeting Nazis Halfway." Maybe you've read it, too.

I appreciate how Solnit frames love of enemy in an invitational way—well, she doesn't use the words "love of enemy," but I think that's what she's getting at. I've returned to her words several times in the past few days, especially these:

> A lot of why the right doesn't "understand" climate change is that climate change tells us everything is connected, everything we do has far-reaching repercussions, and we're responsible for the whole, a message at odds with their idealization of a version of freedom that smells a lot like disconnection and irresponsibility. But also climate denial is the result of fossil fuel companies and the politicians they bought spreading propaganda and lies for profit, and I understand that better than the people who believe it. If half of us believe the earth is flat, we do not make peace by settling on it being halfway between round and flat. Those of us who know it's round will not recruit them through compromise. We all know that you do better bringing people out of delusion by being kind and inviting than by mocking them, but that's inviting them to come over, which is not the same thing as heading in their direction.[2]

Inviting them to come over.

There's a lot to like about that. Among other things, an invitation is not coercive, which opens up the possibility for new (or renewed) connections between people who do not see things the same way. Also, it prescribes kindness, which is good medicine,

---

2. Solnit, "On Not Meeting Nazis Halfway."

especially when the demonization of others ails us. Then again, there may also be a few cracks in the "come over" plan. I'm thinking of its implication that the Inviter has the answers and will impart wisdom to the Invited. I can hear cries of "elitist" from here, and they may not be unfounded.

In the light of the climate crisis, I'm also wondering if "inviting them to come over" lacks a sense of urgency. It seems to me that those who hear our planet's alarms sounding have a responsibility to actively, clearly, and creatively communicate what is happening—and what needs to happen now. Rather than sitting on our front porch and waiting to see who'll drop by, we need to find old and new ways to reach everyone.

Jesus was really good at that. I'm thinking of his call of the first disciples. He does not say to Peter and Andrew, James and John, "Follow me and I'll make you accountants of the Kingdom." Jesus says to the fishermen, "I'll make you fishers of people." Jesus gets through to them.

Isn't this our work, too?

When it comes to building a broad climate action coalition, your idea of working with the "not optimal" toward "imperfect, stepping-stone solutions" is spot on. And yeah, some folks will never join in. But far more will. And together, we will bring our baggage and our best to fight against the undoing of so much life on Earth.

It'll be a messy mosaic, for sure—the only kind of mosaic I know how to make!

∞

Back in 2015, I led a Youth Group service trip to Boston. One day, a group of us went to Emmanuel Church on Newbury Street to assist with a day-long program called Common Art, a ministry of Common Cathedral. Common Art is the brainchild of a formerly homeless man who once attended Common Cathedral services. Several years ago, Common Cathedral's ministers noticed his artistic talent and asked him to imagine a program to create community through the arts. Common Art is what he came up with.

January 14, 2021

So, we spent the morning making art alongside men and women who had clearly seen more than their share of trouble but who had found in one another real sanctuary. At one point, I sat down at the mosaic table and learned the basics. I set out to make a simple mosaic called "Green Mountain Sunrise." Truth is, it looks more like a seascape. And not a very compelling one either. Anyway, I keep it around to remind me that each of us is a little piece of something bigger. Jagged pieces. Shimmering pieces. All of us broken. And when we come together for a common purpose, a higher purpose, sometimes the light hits us just right, and we begin to slowly recover hope, courage, strength. May it be so.

Be well, friend.

Grace and Peace,
Andy

JANUARY 31, 2021

Dear Andy,

Greetings from New Haven, where I'm happily ensconced back in my tiny divinity school apartment on the second floor of Bellamy Hall. I'd forgotten how much I love the view from here. I can see over the rooftops out to East Rock. The flat horizon to the right of East Rock reminds me that the ocean is out there. And with the dogwood outside the window, it feels like a treehouse.

After eleven months in Ripton, lovely as it is there, I'm grateful for the change of view and, even more, for the return to this community. Isolation was taking a toll. Deeper than I knew. It was a nourishing return; my friends had keys to my apartment, and while I was driving down here on Thursday, they were decorating it for me. They filled it with flowers. And a banner that says "Welcome Home." And another banner that's just a string of red hearts. And a plant. And cards with loving words written on them. Some yarn for knitting. And a cheese pizza from Sally's: love, New Haven style. I'm glad I listened to what my heart was telling me and headed south.

One of my feats of unpacking yesterday was to finally locate your last letter. I was glad to see it, although the reminder it offers of the emotional distance we've traveled this month was a bit jolting. Where the first two weeks of January were a horrifying exhibition of the worst of who we are as a country, the second two weeks have offered some welcome glimpses of the best of us. What a balm that Inauguration Day was. How good it felt to hear words of honesty and hopefulness and poetry. Amanda Gorman, wow.

The thing I'm wrestling with—I suspect the thing many of us are wrestling with—is that there's truth in both halves of this month. Your letter speaks truth. And so did all of those words on Inauguration Day. The work for us is to hold both of those things. To not let the healing rhetoric cause amnesia. It can't remain simply

January 31, 2021

rhetoric. We have work to do. And yes, as you say, that includes loving our enemies. I'm feeling as challenged by that as you. I reread my last letter to you, and I can't help but feel like there was a bit of naiveté in my enthusiasm for befriending those with opposing views. It sounded easy back then. But I don't know how to love the people who stormed the Capitol on January 6 either. And of course, you're right, we have to. Much as I would like to treat Jesus' teachings like a menu at a tapas restaurant—I'll take two of the "love your neighbor," an extra helping of loving God, but that's a hard pass on loving my enemies—I know that's not how it works.

Teach me to love, O Lord.

I appreciate where you take that challenge in your letter. Loving doesn't mean meeting my enemies halfway, but it does try to invite them in from the rain. It reaches across the breach with both kindness and resolve. I find it easier to think about loving my enemies if I put that work under the umbrella of "repair."

I've been hearing that word a lot in the weeks since January 6. In President Biden's inauguration speech: "Much to repair. Much to restore. Much to heal. Much to build. And much to gain."[1] In the Rev. William Barber's homily at the prayer service the day after Inauguration Day: "Choose first to repent of the policy sin. Then, repair the breach."[2] And in your letter.

We've got repair on our minds. And I find hope in that. Because the kind of repair that's in the air these days is like your messy mosaic. It isn't trying to put back together what was, it's taking brokenness and making something new. Whole, but in a different way. A better and more beautiful way. It's the kind of repair that recognizes that turning to the past and righting old wrongs is necessary to the work of pushing forward.

It's the repair that Rev. Kelly Brown Douglas talks about when she talks about reparations. She wrote a powerful piece for *Sojourner* about reparations and the Church that ends like this: at "the end of the day, God calls us all to a future where the first are last and the last are first (Matthew 20:16). Such a future does not reflect

---

1. Biden, "Inaugural Address."
2. Barber, "Inaugural Sermon."

a reversal of privilege and penalty. Rather, it is a time when there is no first or last because everyone is treated and respected as the equal child of God that they are. It is left for faith communities to 'repair the breach' between the present injustice and God's just future. Reparations must not simply look back, but most importantly must push forward."[3]

That's the kind of talk that climate change needs more of. Because the work of climate change is both. It's dealing with the legacies of historical actions, and it's charting a course forward. Both are necessary. Our atmosphere, after all, is a repository of history. That troublesome CO_2 tells the story of our past—of the damage done and the damage to be repaired.

It seems to me that bringing the framework of reparations to climate change does a couple of important things. For one, it points us at the systemic root cause of climate change. The word "reparations" speaks the truth that the same trio of sins—white supremacy, economic injustice, domination—lies at the root of enslavement, genocide, poverty, and ecological destruction. It gets us away from siloed thinking—beloved by the status quo—that wants us to believe these are separate problems. But of course, they're not. They grow from the same twisted root.

And the idea of reparations has been at least implicit in discussions of climate change mitigation from the beginning. The idea is simple. Wealthy countries—the Global North—owe something to those off of whom they have profited—the Global South. The Paris Agreement leans in that direction, setting the expectation that developed countries will provide financial resources and other forms of assistance to developing countries.

The other thing that the idea of reparations brings to the problem of climate change is a concrete way of thinking about how to repair equitably. It points to ideas like carbon debt, which goes something like this: we know about how much of the CO_2 in the atmosphere is due to human activity, and we know more or less where that CO_2 came from. Not surprisingly, most of it came from developed countries. And much of the carbon that came from the

---

3. Douglas, "A Christian Call for Reparations."

JANUARY 31, 2021

Global South was a result of deforestation driven by the insatiable appetites of developed countries. Climate colonialism, it's been called. Developed countries haven't been keen on acknowledging that debt. But from an ecological perspective? If we read the atmosphere as a story of history—each molecule of CO_2 a narrative thread in that tale—then it's unambiguous. Most of the CO_2 in the atmosphere belongs to people like you and me: white, wealthy (relative to the rest of the world) Europeans and Americans.

We have a debt problem. And there are multiple ways to think about carbon debt. Ecologically, it makes sense to me that any deviation from a carbon-neutral life is a carbon debt, one owed to planet Earth. And maybe we'll get there eventually. More commonly, these days, people think about carbon debt in comparative terms—about how much carbon each of us emits relative to the global average. That debt is one we owe to each other. To our neighbors. *That* notion of carbon debt starts with the assumption that the atmosphere is a common good. If you make that assumption, it's easy enough to calculate what equity would look like. Take all of the carbon emitted, divide it by global population, and you get each person's share. Multiply that by a country's population, and you can calculate how much carbon a country would emit if it were emitting only its share. If you compare a country's actual emissions to that number, you can estimate carbon debt or credit. It's imperfect, but it's a place to start. A 2016 paper in *Nature Climate Change* listed the top carbon debtors and creditors.[4] The top three creditors are India, Indonesia, and Bangladesh. The top three debtors are the United States, Russia, and Japan.

You know I have a fondness for concepts that have meaning in both the ecological realm and the theological realm. I think that the new stories we need to tell will require such words, and the concept of debt is just such a thing. It's a quantifiable scientific data point. Ecologically real. And it's also moral. Unlike personal financial debt, our debt problem here in the Global North harms other people. It would be as if I overdrew my checking account and someone on the other side of the world was taken off to debtor's prison.

4. Matthews, "Quantifying Historical Climate and Carbon Debts Among Nations," 60–64.

January 31, 2021

From the perspective of the earth's carbon cycle, I am a debtor to most of the rest of the human beings alive today. That's a different way of telling the climate change story. I think it's a powerful one. A reparations framework helps us to grapple with how to respond to the reality that the privileged are really debtors. Paying back that debt would look like reducing carbon emissions. Strengthening carbon sinks. Repairing the harm done by our debt would look like sharing technologies and financial resources to allow developing countries—our creditors—to mitigate the damage we have done to them.

In the sermon I preached after the inauguration—on the calling of the disciples in Mark's Gospel—I talked about Jesus calling us into a new story. His story. I said that to step into that story, we needed to leave old stories behind. The stories that led us to this perilous place, this tipping point moment. The stories of white supremacy. Of nature dominated, objectified. Of Earth without limits. The stories that maroon us, islands each, in the lonely fallacy of individualism. And as I said in that sermon, leaving old stories behind is hard. It's less like dropping a fishing net and more like untangling ourselves from a whole-body knot. I think the idea of reparations is integral to the work of letting go of the old stories.

Reparations might be the lever we need to move us from the path we're on—our old story—to that new path that the future needs us to be on. The new story. The one that narrates the path toward justice.

Zoom calls, so I'll draw this to a close. Say hi to the loveliness of Vermont for me. I miss you all up there. It's a strange feeling knowing that the easy, four-hour drive back up to VT is not really an option in these pandemic days. Dare I hope that spring may eventually allow for easier inter-state visits? I do hope.

And in the meantime, friend, be well.
Andi

FEBRUARY 24, 2021

Dear Andi,

The tale of your return to New Haven tells me you've got some people down there who think you're better than all right. They welcomed you with Sally's pizza, for heaven's sake. Other stuff too—but I remember the pizza. And I do remember it. How could I forget Sally's? Beautifully fruity. Salty meets sweet. Just the right amount of soot on the crust, on your fingers. All other Neapolitan pizza is measured by it.

One late spring night, more than twenty years ago, Gwen and I and some div school friends, Eric and Rachael Bowman, watched a Sally's special go by our table as we were enjoying the pizza we had ordered. We inquired. It was called The Primavera. We ordered one. And *that* pizza—that was the single best pizza I have ever had. A celebration of the season's first harvest with people I love.

Anyway, thank you for invoking the culinary wonders of Wooster Street. It made me a little hungry, but more than anything it made me happy. Good people help—O Lord, do they ever. We call them saints. Your "welcome home" story includes a few. Thanks be to God.

So, I read that last letter of yours when it arrived a few weeks ago. Then I got an early start on my Lenten journey. My parents were laid low by Covid—"We're feeling much better," says my mom through a detectable wheeze. Gwen's mom returned home for less than a week after two months in the hospital and nursing home—she's back at Helen Porter. Our dog Teddy was diagnosed with inoperable bladder cancer—Rachael adds his name to our table grace each evening. And just three days ago, I drove the two miles to Middlebury College with Mary, then I drove home alone. I know she isn't far, but she isn't home. I miss her. It's her birthday today, the first day of the Spring semester.

February 24, 2021

I reread your letter this morning and was buoyed by your mention of "repair." Repair tells a truth I need to hear: brokenness abounds and we are equipped to do something about it. Repair is a good word for times like these. It's honest about "the twisted root of enslavement, genocide, poverty and ecological destruction" that continues to bear the most bitter fruit. And it gives me hope to remember that individuals and communities can heal and grow stronger.

There's a lot of repair work to do these days. And for what it's worth, I think the idea of dominion deserves to be near the top of the service slip. Dominion—*radah* in Hebrew—is a wreck, and we can do something about that.

The other night I took Teddy outside to pee. We walked down to the Hurd Grassland, and I think we were both surprised: the stars were weirdly dazzling. Teddy noticed. He looked up a few times. He was more distracted than usual.

The next afternoon, a sky-smart neighbor told me I might have seen two thousand stars the night before. It sounded like a lot, until he said there's something like a hundred billion stars in the Milky Way, and something like a hundred million galaxies in the universe. I looked it up later to be sure I heard him right, and yes, the universe is home to some three hundred sextillion stars[1]—a 3 followed by twenty-three 0s. That's some neighborhood—a Psalm 8 kind of place.

> O Lord, our Sovereign,
> how majestic is your name in all the earth! . . .
> When I look at your heavens, the work of your fingers,
> the moon and stars that you have established,
> what are human beings that you are mindful of them,
> mortals that you care for them?
> (Psalm 8:1–4, NRSV)

I love that Psalm 8 props up our small selves before the vastness of it all. But I struggle with the rest of the psalm—the part about us being "a little lower than God" and "crowned with glory

---

1. Memmott, "'Trillions Of Earths.'"

and honor"—the part about us having "dominion over" the creatures of land, sky, and sea.

The question I've been wrestling with lately is not whether we are highly and mightily powerful. We have made it very clear that we are capable of exercising power over other creatures and other humans. In fact, "carbon debt" and "white supremacy" would be nonsensical terms if dominion-thinking weren't a thing. It's definitely still a thing.

The question ricocheting inside me is, why would humans exercise dominion in ways that burn the house down? What good is power if it ruins beauty? What kind of Kingdom is that? What kind of republic? And even if "the earth is not our home," it's beyond me how anyone could expect to enjoy the view from heaven after spoiling this paradise of God's making. Theologically speaking, dominion is badly lost if it lingers anywhere near the exploitation of others. The prophet Ezekiel calls out Israel's lost leaders for exactly that reason. Where "shepherds" is code for king and "sheep" are the people of Israel, the prophet is piercingly clear in his assessment of the royal dominators:

> Ah, you shepherds of Israel who have been feeding yourselves! Should not shepherds feed the sheep? You eat the fat, you clothe yourselves with the wool, you slaughter the fatlings; but you do not feed the sheep. You have not strengthened the weak, you have not healed the sick, you have not bound up the injured, you have not brought back the strayed, you have not sought the lost, but with force and harshness you have ruled them. So they were scattered, because there was no shepherd; and scattered, they became food for all the wild animals. My sheep were scattered, they wandered over all the mountains and on every high hill; my sheep were scattered over all the face of the earth, with no one to search or seek for them.
> (Ezekiel 34:2–6, NRSV)

Dominion's been a wreck for a while.

Still, we can choose to steer our power in the opposite direction. We can choose to use our God-given power in ways that honor the goodness of creation. It is possible to lead with caring, especially

February 24, 2021

for those who do not have what "the haves" have. It's possible to read "dominion" and think "caretakers of this planet and all its species." It's possible to hear dominion as a call to tend the Garden, so that we and our companion creatures can live now and in the future. It's possible because that's what *radah* means.

It's there in Psalm 72. The king in this psalm does things differently:

> He delivers the needy when they call,
> the poor and those who have no helper.
> He has pity on the weak and the needy,
> and saves the lives of the needy.
> From oppression and violence he redeems their life;
> and precious is their blood in his sight.
> (Psalm 72:12–14, NRSV)

This passage provides a glimpse of dominion in a political context, but this dominion-mindset stretches into the ecological realm, too. Ellen Davis, a brilliant and beloved professor at YDS during my time there, tells it like it is: "We fulfill our role in the created order only when we recognize our responsibility to help perpetuate other creatures' fruitfulness."[2] Responsibility—yes. To repair *radah* is to reclaim our responsibility to care for our common home. This is not Ellen Davis's insight alone. The same notes are struck in Eugene Peterson's translation of the end of the opening chapter of Genesis.

> God blessed them:
> "Prosper!
> Reproduce!
> Fill the earth!
> Take charge!
> Be responsible."[3]

There it is again—responsibility.

It's also there in Pope Francis's second encyclical, *Laudato Si'*. The leader of 1.2 billion Catholics tries to repair the common misunderstanding of dominion because it "has encouraged the unbridled exploitation of nature by painting [humans] as domineering

2. Davis, "Meaning of Dominion."
3. Peterson, *The Message*, 4.

and destructive by nature. This is not a correct interpretation of the Bible as understood by the Church."[4] The Pope goes on to say, "We must forcefully reject the notion that our being created in God's image and given dominion over the earth justifies absolute domination over other creatures. . . . Each community can take from the bounty of the earth whatever it needs for subsistence, but it also has the duty to protect the earth and to ensure its fruitfulness for coming generations."[5]

Protect the earth. Ensure its fruitfulness. Act responsibly. There are a ton of us who want to help mend what's torn. And I'm foolish enough to believe that when we figure out how to pool our God-given power, repair will beget repair over all the earth. This is my hope for dominion.

I wish you a blessed walk through these holy days.

Peace,
Andy

---

4. Pope Francis, *Encyclical on Climate Change & Inequality*, 43.
5. Pope Francis, *Encyclical on Climate Change & Inequality*, 44.

MARCH 5, 2021

Dear Andy,

My journal tells me that it's day seventeen of Lent. My heart is mounting a pretty solid counter-argument that we're somewhere around day 374. It sounds like your life has been charting a similar course of late.

And it sounds like we've both reached that age at which the reality of aging parents gets real. And Covid doesn't help. Not one bit. I read that added layer in the path of worry you and Gwen have been walking lately with her mom and your folks. I feel it myself, too. Late last week, my father emailed me. He wanted to spend some money—a perfectly reasonable expense, and he's got plenty of it to spend. But this time, his confidence in his ability to do things wavering, he wanted reassurance. Negotiating that role-reversal—the erosion of that parental bulwark—is never easy. Negotiating it from afar, over a screen? Hard.

So, I did what I do when "hard" starts tilting toward "too hard" and went for a walk. Motion helps. Spring helps. And spring's here in New Haven. My walk took me past a hillside garden on Saint Ronan Street resplendent with crocuses, snowdrops, and winter aconite, all in bloom. I remembered that it was almost exactly a year ago when I stopped at precisely that spot and smiled at those flowers. I remember thinking to myself: how wonderful that I get to experience a Connecticut spring for the first time in thirty-five years! Forty-eight hours later, I was back in wintry Ripton and the world was shutting down. My heart didn't quite know what to do with the realization that we've been in this wilderness now for a full year. Lean into joy at the flowers' promise of a second chance to experience my first Connecticut spring in thirty-five years? Or give in to grief at this anniversary of when the world changed forever? Like I said: day 374 of Lent.

March 5, 2021

I came in from that walk and picked up your letter again. Our letters buoy me, too. It's not just the content, although there's plenty in your last letter to lift my spirits. It's also the way they remind me that my insufficiency to fix any of this—climate change, life, death—is OK. We're all necessary, but none of us is sufficient on our own. And the good news is, we're not on our own. Any of us. Over and over again, our letters remind me of that. My voice needs your voice. Our voices need other voices. All of us need God.

That train of thought delivered me to the last paragraph of your letter—to your beautifully "foolish" belief that repair will beget repair. Provided, that is, we figure out how to pool our God-given gifts. That sounds right to me. And it set me off on a project to read about climate governance. The Day Missions Room is my preferred venue for reading these days. Covidtide has left the room mostly empty. Tables are limited to one person. The prime seating—those balcony tables—is closed. But with the morning sun streaming in, the shelves of old books, the hush: it's a good place to go deep. I spread articles and books out on one of those big, old wooden tables and started looking for an answer to your question. How do we pool our God-given gifts?

Polycentric climate governance. That's one answer I found.

A mouthful of an idea. It goes something like this.[1] The old way of solving international problems involved agreements between states. National governments were pretty much the sole actors in that drama. Polycentric governance recognizes that there are many different actors working toward solutions: nations, regional and local governments, private companies, NGOs, churches, individuals. Those actors work at different scales using different tools, but they're all striving to right the same wrong. Climate change, in this case.

Polycentric climate governance recognizes that multiplicity of actors. It acknowledges it, embraces it, and gives attention to how to integrate those different actors. How to pool their gifts. Advocates of polycentric governance extol its many benefits: learning, experimentation, diversity, participation, connectivity. It's a

---

1. Schoon et al., "Principle 7," 226–50.

pretty great list of words, isn't it? And sounds an awful lot, to my ears, like the kind of dominion you were describing. I especially love the humility inherent in a model of governance that embraces learning and experimentation: one group trying something, learning, sharing. Other groups listening and learning from them. That part reminded me of what you were telling me the other day about Gwen's work on the Weybridge Energy Committee and the benefits that would come of sharing that group's success more broadly. Polycentric governance says "Yes! Let's learn from each other." And it's inclusive. Everybody has a seat at the polycentric table. It promotes diversity—in policy, in approach, and in the human beings doing the work. And it enables more participation, by more people with more different kinds of gifts. Repair begets repair. That's what I hear.

Where your question about pooling really comes into focus is around that word "connectivity." Polycentric governance requires forging intentional connection between different actors. Building trust. Flexing the muscles of cooperation. That cooperation can take different forms—from simple communication links to formal partnerships to joint projects.[2] The connectivity at the heart of the idea of polycentric climate governance is all about pooling our gifts.

Good things happen when we do. There's a beautiful example—a textbook case of polycentric governance—to be found down on the southern border, in Arizona. The Malpai Borderlands Group started about thirty years ago. I remember hearing about it when I was in grad school in Tucson—it was just getting started then. I remember thinking it sounded revolutionary. And it is. The group's an unlikely alliance of ranchers, environmental groups, and local governments—folks who tend to sit on different sides of a decidedly un-round table elsewhere in the West. They came together when a group of developers set their sights on some open land in their shared desert-grassland neighborhood.

In an article about them in *High Country News* back in 2015—with the fantastic title "Why being a good neighbor is a good idea"—the executive director explains their success in simple terms. "You start with something you agree on instead of something you

---

2. Briggs, et al., eds., *Principles for Building Resilience*.

disagree on."[3] And so, they did. They agreed about the common ethos that guided their lives. *Neighboring*, they call it: a responsibility to take care of each other. They agreed that they loved the land. As it is, undeveloped. So, this disparate group of entities, organizations, and neighbors formed the Malpai Borderlands Group. It's now an ecological—and community—success story and an example of what polycentric governance can achieve. And these days, it's happening all over the place.

The United Nations has a website to catalog the different climate actions being taken by different actors.[4] When I last checked, it had 18,553 actors around the world, representing 27,493 discrete climate actions. Some are big, like the Small Island Developing States (SIDS) Lighthouse Initiative:[5] thirty-six small island states from around the world with thirty partners—developed countries, private companies, NGOs. The initiative's goal is to help the SIDS—among the world's most vulnerable to climate change—move rapidly to renewable energy. Between 2014 and 2019, renewable energy installations among the member states increased from about 97 MW to 1,180 MW. That's still a small number—by comparison, the US was on track to add something like 27 GW (27,000 MW) of wind and solar in 2021. But it's a rapid increase in a part of the world with limited resources. Other projects are small: like Campeche, a state in southeastern Mexico that has committed to a 50 percent reduction in regional emissions from agriculture, forestry, and land use by 2030.[6] A small region doing what it can. One group of people pooling their God-given gifts to contribute to planetary good. As I was exploring the UN's website, I found myself wondering if we could push the conversation even further. What if we didn't limit ourselves to people pooling their God-given gifts? What if we began to act as if we were in partnership with all of creation?

That question has been in the back of my mind since the fall, a seed planted in Professor Willie Jennings's class on Natural

3. Mockenhaupt, "Why Being a Good Neighbor Is a Good Idea."
4. Climate Analytics and NewClimate Institute, "Climate Action Tracker."
5. International Renewable Energy Agency, "SIDS—Small Island Developing States."
6. Climate & Clean Air Coalition, "Mexico."

March 5, 2021

Theology. Professor Jennings challenged us to consider what it would mean for our relationships with God and the world if we took the rest of creation seriously. If we engaged our non-human neighbors as living things with agency. With their own relationships with God. Knowing things we don't know. We read some smart people making some solid arguments that Christianity is rooted in that soil. Ecology is, too.

To take that idea seriously is to step into a very, very different way of interacting with the world—one that is, interestingly, both consistent with and in tension with the idea of dominion you invited into our conversation in your last letter. Consistent because a world of creatures with agency, with their own relationships with God, is a world that invites power *with*, not power *over*. In tension, because such a world is harder to fit into the hierarchy implied by dominion. That's what I hear you struggling with in Psalm 8. Me, too.

That struggle aside, an understanding of the world that takes nature seriously leads us to acknowledge that there's wisdom in creation. A wisdom not our own. That's a truth I can get behind ecologically and theologically. When I think about the connections among the centers that make up polycentric governance, I start imagining an ecological template on which it's constructed. I start imagining how the voices of non-human creation might guide us if we began to see creation as partner, bringing gifts that can be pooled with ours.

I see traces of that listening in the work of the Malpai Borderlands Group. I think the people in that group heard the wisdom of the land they loved. They understood that the big, open spaces that make their hearts sing—the ones that developers wanted to turn into ranchettes—were not just beautiful, but necessary. They understood that biological complexity—including the ridge-nosed rattlesnake and the Chiricahua leopard frog they saved from extinction—matters. For its own sake and for theirs. The natural world was partner in that work. A mostly silent partner, maybe, but a partner whose needs were allowed to be part of the conversation.

I think we need more of that. The kind of listening, respect, inclusiveness that builds connections bigger than our human lives

March 5, 2021

and human institutions. And like you, I'm plenty foolish enough to believe that we can get there. All creation, working together. Repair begetting repair. May it be so.

Be well, my friend. Give my love to your people and to my beloved Vermont.

Peace,
Andi

MARCH 28, 2021
PALM SUNDAY

Dear Andi,

I hope you're well. Thanks for the updates on your dad's condition over the past few weeks. Your field-note emails from a Florida hospital have much to say about what a heart can hold. How are we to care for a loved one who is slipping into the mist of memory loss? For what do we pray? What is it that draws us back to the bedside? What is it that sends us outside for decidedly different air? I'm sorry for all the weight you're bearing, friend. God be with you all.

Here on Meadow Lane we're still getting used to being a table of three, but Ella and Mary are doing well. The reports from Davidson are positive: Ella's excited to jump into her thesis next year. She's going to look at the experiences of the first women to matriculate at several formerly all-male colleges and universities in the late sixties and early seventies. And Mary is off to a good start at Midd. I saw her at College Vespers on Thursday night. She seems solid, sounds like herself.

Last week, Mary texted Gwen and me a slide from her Environmental Science class. The slide's many shades of red, purple, blue, and gray were telling a story about the land around The Knoll. I followed up with a phone call, and Mary started talking about photosynthesis: the slide shows a spring scene—the darkest reds are way ahead of the blues in converting sunlight into sweetness—the stands of pines that look like bunches of purple grapes have a photosynthetic party before the other trees leaf out—the blue blotches are fallow fields, where new grasses have not yet surpassed the old. I love when Mary gets talky.

Anyway, I can't keep my eyes off that slide. It's fit for framing. It looks like a huge autumn-colored carpet. Or maybe a fine arts

March 28, 2021 Palm Sunday

drone shot, like Caleb Kenna takes. And it's all the more beautiful because I walk those border woods and fields a lot. That place matters to me: the Dutch Belts on the Scholten's dairy farm, the white pine forest, the wetlands, the haymaking fields, the college's organic garden. The eastern edge of that open space is within sight, as I write. I'm used to thinking of it as one distinct area, but Mary's slide shows me how many ecological neighborhoods those hundreds of acres support. The word you use to introduce polycentric climate governance—multiplicity—is close to what that photosynthesis slide communicates in color.

I knew exactly nothing about polycentric climate governance before reading your letter. At first pass, I'm taken with the idea. I like the sound of a whole cast of characters working on manifold solutions to climate change. Forging partnerships. Sharing best practices. Learning from each other. The markings of this many-centered approach—especially the diversity of the actors and their hoped-for chemistry on stage—remind me of how you speak and write about the interdependencies written into ecology's script. You have made it clear to me and our students that ecology is a story of profound connection, so it makes sense that climate change solutions will continue to arise from the discovery and development of networks of people and organizations—and yes, creation itself. It's a compelling framework for pooling our God-given gifts and sharing the stories we know.

The one you tell about the Malpai Borderlands Group is a gift in itself. Thank you for that. It's a raucous celebration of collaboration. Ranchers hanging out with tree huggers? Tree huggers hanging out with politicians? It sounds like the kind of round table alliance that we're after, doesn't it? Anyway, my hope swells whenever unlikely partners work together for the common good, and as far as I'm concerned, those happenings in southern Arizona need to be made famous. Who's going to write *that* script?

The number of climate action groups around the world offers ample encouragement. The UN's catalog of climate actions is hefty. After reading your letter, I felt inspired to do some sleuthing of my own. I was curious to know if there was a directory of Christian or faith-based groups committed to the care of creation and the

March 28, 2021 Palm Sunday

work of addressing climate change. I didn't find a comprehensive list, but I did see dozens of compilations. It was good to run into some old friends—Earth Ministry, Au Sable Institute, Interfaith Power and Light, Web of Creation. Still, how many congregational Green Teams are out there? How many small, religiously affiliated groups exist alongside Addison County's Interfaith Climate Action (ICAN)? How many houses of worship have invested in solar arrays? How many church-based farmers markets support sustainable agriculture, like the one Gwen started at Spring Glen Church in Hamden, Connecticut in the early 2000s? I don't know, but I have a hunch that it's a very large and ever-increasing number. Maybe there's someone out there who has the time and mind to create such a database. It would be a useful tool to link groups of people working on similar projects across the country and around the world.

One of the most persuasive Christian climate action groups I know is Young Evangelicals for Climate Action (YECA). They hold sway. For the past decade or so, their good work has mobilized waves of youth and young adults, influenced big-time evangelical leaders, and held politicians accountable. They take action and make environmental justice a priority. This is from the News & Notes section of YECA's website, dated March 22, 2021:

> Last week, Rep. McEachin (D-VA), Rep. Grijalva (D-AZ), and Senator Duckworth (D-IL) introduced the Environmental Justice for All Act. In response, YECA National Organizer and Spokesperson Tori Goebel released the following statement:
>
> As followers of Christ, we must boldly address and uproot historic and ongoing injustices in the United States. The impacts of corporate pollution and climate change disproportionately impact Tribal and Indigenous communities, communities of color, and low-income communities. It is well past time to confront this reality and strive for a more just world in which all of God's children have access to clean air and safe water. We cannot solve the climate crisis without caring for our neighbors and addressing environmental injustice, and that is why

March 28, 2021 Palm Sunday

> YECA applauds the introduction of the Environmental Justice for All Act.
>
> For far too long, Tribal and Indigenous communities, communities of color, and low-income communities have been excluded from the decision-making process. This powerful, comprehensive legislation was developed through collaboration with the communities most impacted by environmental racism and oppression and it seeks to alleviate the ongoing burdens caused by toxic pollution and environmental hazards.[1]

This is a good word—and it's something that could have been authored by a team of leaders in the United Church of Christ, from the other side of the theological spectrum! Like many progressive Christians, YECA wants to add more chairs at the table of environmental justice and climate activism. More stakeholders. More voices, especially the ones long silenced. More attention paid to the needs of sinfully dismissed communities. YECA's strong affirmation of collaboration and their insistence on linking climate action to justice for all are ties that bind Christians across denominational lines.

And there's something about this that brings the Holy Spirit to mind. I'm thinking of the story of Pentecost (Acts 2). As the story is told, the breath/wind/spirit of God creates Christian community. A variety of people from all over, with their different hang-ups and hopes, are brought together to be the body of Christ in the world. And really, this is all the permission I need to imagine God prepping unlikely partners for the work of pushing toward the good of all creation and against the tide of the climate crisis. Forging partnerships. Sharing best practices. Learning from each other. In the reaches of the Christian imagination, there is nothing more polycentric than the Holy Spirit. In this, I find reason to hope.

I'll sign off here. My surgery went well on Friday, but my body is telling me it's time to rest. May your walk through Holy Week bring you to the awesome edge of sacred mystery and to the

---

1. Young Evangelicals for Climate Action, "YECA Supports Environmental Justice for All Act."

March 28, 2021 Palm Sunday

assurance that nothing, not even death, will be able to separate us from the love of God in Christ Jesus our Lord.

A blessed Palm Sunday to you, friend.

Peace,
Andy

APRIL 8, 2021

Dear Andy,

By the time this letter arrives in your mailbox, you will be nearly at the start of that long-overdue sabbatical. The thought makes me glad. I pray that the months to come bring you regeneration and renewal. Rest and healing. Long walks in the fields and forests. Joyful times with the family. The stillness of a good silent retreat. May that deep well of yours refill, dear friend.

Speaking of things that refill us, I got a sweet note from Mary on Easter. She's a kind-hearted soul, that daughter of yours. She reported that the Nagy-Bensons—and the Bensons, thanks to vaccinations!—had a good Easter celebration. I was glad to hear that. The Easter service was lovely, start to finish. It occurred to me, as I was stitching it together, how much we've learned about virtual worship over the last year. We've gotten good at it. It also occurred to me to be glad that I didn't know, a year ago, that we had two Easters of pandemic to come. I don't think I could have handled that knowledge. Anyway, Easter down in New Haven was wonderful. After watching our Easter service, I gathered for brunch with friends. The prayer spoken over quiche and salad felt like an extension of worship. Friends, food, prayer: it felt like holy time and a welcome respite amidst this challenging march of days.

And then Monday arrived. Monday was the day my dad was transported to Connecticut. He made it in one piece, but the trip was hard on him. I saw him briefly, in the parking lot outside the Caleb Hitchcock Health Center at Duncaster, before he headed inside for two weeks of quarantine. There was no recognition in his eyes. Just wild-eyed fear. I'm having trouble shaking the memory of how it felt to look into familiar eyes that had become an abyss.

Prayer helps. And I was noticing yesterday that my prayers have made their way along that hallway of hope that you sketched out for our class so long ago. Prayers for a full recovery gave way to prayers for enough recovery. Enough that he can enjoy watching a baseball game on TV. Or music. The presence of his family. Monday sent me further down the hallway. I'm praying now for peace. What else can I do?

Somehow, in between all of the heaviness that life holds these days, the semester continues. And the life of the church in Hinesburg continues. Sometimes, I manage to focus my heart and mind on those things. I'm learning to be gentle with myself in the times when such focus eludes me. The one thing I can reliably do is sermon-writing. It grounds me. And the insight with which your letter ends—the bridge you built between the word "polycentric" and the Holy Spirit—stayed with me as I pivoted into sermon-writing this past weekend.

In addition to my sermon for this coming Sunday, I had to write a homily for my preaching class. My text was Acts 4:32–35—that lovely, post-Pentecost moment in the life of the church. "Now the whole group of those who believed were of one heart and soul," it begins. I read it with your letter in mind. Your insight prompted me to read "one heart and soul" as a statement of common cause, not individual uniformity: a diverse group of people living grace-motored lives in pursuit of the common good. That's what I hear. I followed your lead and took the homily straight from there to polycentric climate governance. (Do you suppose that's the first time the phrase "polycentric climate governance" has been used in a homily?)

As you probably remember from your days as a teacher of preaching at YDS, the preaching moment is followed by feedback from the instructor and the class. One of my classmates offered that the homily made him wonder how the world would be different if scientists learned to tell stories like preachers do. Indeed. Another noted that the story of interdependence that ran through it brought her into the homily in a way that other stories might not have. Those two comments, together, have left me wondering about the stories I choose to tell. I love to tell the story of interdependence. And oh, how I love the phrase "common good." What about the stories I'm not telling? And who am I missing with the stories I am telling?

APRIL 8, 2021

Who is it that would have heard my homily and felt unconvinced? And why?

My search for answers to that question started close to home, with the Yale Program on Climate Change Communication. They're the ones who added crucial nuance to the dichotomous division of Americans into "people who believe in climate change" and "people who don't." America, they found, can be described by not two but six categories. The Six Americas: alarmed, concerned, cautious, disengaged, doubtful, dismissive.[1] I think it's Katharine Hayhoe who has said that she doesn't waste her time on the dismissives.[2] But the cautious, disengaged, and doubtful? There's room there for conversation. So, that helped me refine my question. How might the stories I tell—the ones that center ideas like "the common good" and "interdependence"—land with those folks?

This program regularly polls Americans to find out what we think about climate change, at a really granular level.[3] The granular details tell a more complicated—and to my eyes, more hopeful—story about this messy, divided country of ours. If you ask people whether they think individual citizens should do more to address climate change, most say yes. There are some pockets of skepticism, but the majority of Americans—64 percent of them—think that individual Americans should do more. A bunch are undecided. Only 11 percent think we should do less.

But if you ask people whether their *governor* should do more, a different story emerges. There's far less support and a much clearer geographic divide. Americans' opinions about government-driven solutions follow the bi-coastal pattern we've seen on electoral maps. That pair of patterns tells me that the middle of the country is full of people who believe that individual citizens ought to be doing more to address climate change, but that government shouldn't. So, if I were preaching my homily to those folks, how would my story of the "common good" land? I'm guessing it would land like a call

1. Yale Program on Climate Change Communication, "Global Warming's Six Americas."

2. Reubold, "Katharine Hayhoe: Bridging the Climate Change Divide."

3. Yale Program on Climate Change Communication, "Yale Climate Opinion Maps 2020."

for big government. So, now I'm wondering: how can I tell the story that I need to tell without pushing those buttons? Is it possible?

One more data point. Ask people whether corporations should do more, and fully 70 percent of Americans think that corporations should do more to address climate change. That's a big enough number that it has to include a whole lot of strongly pro-capitalist, free-market-true-believers. And yet, some of them are comfortable with asking corporations to do more to address climate change. I'm fascinated.

When I put those three data points together, they start to point me toward a more complicated story about Americans and climate change. One that suggests that someone could reject the premise of my homily not because they didn't believe in climate change, but because "common good" isn't their thing. And that got me thinking about Dan Kahan's work. Kahan's a professor at Yale Law School who does really interesting work on cultural cognition. In 2007, he wrote a paper on something called the "white male effect."[4] In a nutshell, studies have shown that white men, on average, tend to assess risk differently than other groups. Presented with a common threat, they will evaluate it as less of a risk than women or African Americans. Kahan's insight, back in 2007, was that the pattern could be attributed to a particular worldview. The people who tested as "fearless" (i.e., saw a hazard as less risky than other folks) were mostly white and mostly men, yes. But more than that: they valued hierarchy and individualism. They did not value egalitarian or communitarian principles.

And I wonder if that's part of what we're seeing in those maps: the individualistic, "pull yourself up by your bootstraps" American ethos. An ethos that is likely to minimize the risk of climate change, for starters, but is also likely to say that if climate change is a problem, we fix it with individual actions, not collective actions. Now, I'm convinced that we need collective action. But might that insight—if I'm putting the pieces together the right way—point to a way to expand that polycentric table to include folks who are not going to say "yes" to an evening of conversation about the common

---

4. Kahan et al., "Culture and Identity-Protective Cognition," 465–505.

April 8, 2021

good? Is there a way to build a bridge from the common good into a more individualistic perspective that will hear "common good" as code for globalism, communism, and all other manner of evils? Or are those inherently at odds with one another?

That question sent me back to the Malpai borderlands. Knowing what I do about rural Arizona, I'd guess that the density of white men who prize individualism is pretty high there. And yet, they worked together, for the common good, as I hear it. So, how'd that happen? I went back to the *High Country News* profile of the Malpai Borderlands Group and found that word I mentioned in my last letter. *Neighboring*. Turns out that when I sent it to you, I bent it to my own communitarian worldview. That's not how the *High Country News* described it:

> The Malpai Borderlands Group has formalized a particular western trait that has long defined daily life around here. "Neighboring," some call it, a way of giving others their privacy while remaining available in case they need you. The notion captures a kind of frontier ideal, an acceptance of the individual's autonomy and self-reliance, tempered by recognition of the precarious and occasionally dangerous nature of outdoor work and the environment.[5]

Rereading that paragraph was an "aha" moment for me. A simple word like "neighbor" can contain multitudes! Neighboring in the Malpai Borderlands sense recognizes the need for interdependence—but between individual actors. The idea of the common good, on the other hand, includes an implied commitment to a purpose that might genuinely be at odds with individual desires. And an implied commitment to prioritize that bigger, shared purpose.

All of which is to say, the answer to my question about different stories is not an answer about semantics alone. Those maps make that clear: some of us believe that the government ought to be acting on climate change and some of us don't. But the maps also suggest that we may differ more about *how* we go about addressing climate change than we do about *whether* to address it. I see hope there.

---

5. Mockenhaupt, "Why Being a Good Neighbor Is a Good Idea."

APRIL 8, 2021

As you say at the end of your last letter, the pathway out of this mess we're in will mean forging partnerships. Unlikely ones, sometimes. And that means figuring out how to talk to each other. I think it means telling more different kinds of stories. We need the stories that will help us connect people and groups that see the world really differently. Bridge stories.

I feel like the preacher in you is going to have a thing or two to say about that. I look forward to hearing it. In the meantime, wishing you and the family a happy mud season. Be well, friend.

Andi

P.S. The birds here in New Haven apparently detect sunrise better than human eyes. The cardinals and robins that hang out in the maple tree outside my window start singing at about 4:00 a.m. I can't be anything but joyful to be woken up in the pre-dawn dark by birdsong. At best, I manage 30 seconds of mild crankiness before I smile into the darkness. So, I woke up early this morning with your photosynthesis map on my mind. It's been nagging at the edge of my consciousness for a while now.

In the letter I mailed yesterday, I wrote a sentence about my aha moment—the one where I realized that word "neighbor" contains multitudes. It occurred to me this morning that that's what the photosynthesis photograph says, too, in a way. That's another way of talking about the multiplicity you describe in your letter. Ecosystems are all about neighbors, really. It wouldn't be wrong to say that ecology is the science of neighbors. But within that word—neighbors—oh, goodness: there are so many variations on that theme. Variations that reflect things like, how far away are your neighbors? Are they going after the same food you are? Do they help you—like the fungal symbionts on the roots of the pine trees—or hinder you—like the buckthorn in the understory of those pine forests, crowding out native species? Are they the same species—like the grasses in the hayfield—or different—like in those wet meadows? Do you interact with them all the time—tree roots interconnected by threads of mycorrhizae—or just occasionally— like pollinators that come in the spring? The deceptively simple

April 8, 2021

reality of interdependence hides within it a multitude of kinds of neighbors. That's part of the multiplicity that your eye (and mine) finds so pleasing about the photograph. And gosh, if we zoomed out? The multitudinous nature of neighboring would become vaster still. We'd see the ways that neighboring looks different in the desert—mostly, neighbors are far away so that they don't compete for water—or in the rainforest—where neighbors just grow all on top of each other. We humans like to think we're different from other creatures, but . . . maybe we're not.

Ecologically speaking, everyone has neighbors—and we all need our neighbors—but that means a million different things. Neighbor is an expansive word. And so, I wonder if another way of saying what I was trying to say yesterday is that if I am to build bridges, to talk about climate change in words that are invitational to people with whom I disagree about a lot of things, I need to find expansive words. Words that can contain a multiplicity of meanings. Those words, in my imagination, become the decking of those communication bridges that are the dotted lines on that polycentric governance diagram. They're the ones that allow us to walk toward each other. To find our unity of purpose.

And I'd hazard a guess that some words have spaciousness within them and others don't. Some words are built for precision. They're built to mean one thing. I kind of feel that way about "common good." If we try to make it spacious, it loses its force—its power to change us. Its power lies in its lack of ambiguity. Common good: it's uncompromising. And so, it holds us accountable. So maybe we don't start there. Maybe we build *toward* those words.

Anyway, that's what this morning of foggy skies and birdsong brought me. If there's any insight here? Well, then, thanks be to God for that.

APRIL 28, 2021

Dear Andi,

How you manage a full load of classes and Marquand Chapel duties (in New Haven), and an internship (in Vermont and online), and your dad's care (now in Bloomfield), especially that, along with a whole host of pandemic-shaped "ands," is beyond me. I pray for your dad. I also pray for you. I hope you find a clearing to set down the heavy stuff you've been carrying. Bless you, friend.

Also, thank you for sending me into sabbatical with a benediction. I appreciate that. Today is day twelve of this four-month time apart, which means I have just begun powering down, like a pilot at the end of a six-year flight.

At the moment, my mind's still more than half full with the people at church. Like you, I heard about Jim Blair's death early Sunday morning. So, I'm thinking of Elise. I'm thinking of my visit with them a few weeks ago—the gratitude and the contingent farewells. I'm thinking of the book of his *National Geographic* photographs on our coffee table. I'm thinking about Elizabeth and the holy work before her. I'm thinking about not being there for them. And I'm thinking that if I were to return from sabbatical for this funeral, then . . .

My head knows, but my heart aches.

So, the letting go might take some time, but as I start to catch up on sleep, and the fog begins to lift a bit, I'm grateful for this extended sabbath. And I keep thinking about time, about how we talk about it in terms of telling. We learn how to tell time, which can sound like we have something to say to time. And perhaps we do—slow down, *speed up!* But what I'm wondering is what these sabbatical days will tell me.

This is what I can tell you so far. The other day, Gwen, Rachael, and I took a weekday walk on the Trail Around Middlebury

April 28, 2021

in search of woodland flowers. We strolled alongside carpets of trout lilies and past drops of hepatica and bloodroot on the western slope above Otter Creek. The lightest yellow-green hues of early spring left us no choice but to pull "Nothing Gold Can Stay" from the shelves of memory. And when we returned home, I sat in a folding lawn chair beside the chicken coop and our red wheelbarrow, and I laughed at the realization that I had stepped unwittingly into William Carlos Williams's masterpiece. So, that night, I put those poems beside each other—both from 1923.

Nothing Gold Can Stay

Nature's first green is gold,
Her hardest hue to hold.
Her early leaf's a flower;
But only so an hour.
Then leaf subsides to leaf.
So Eden sank to grief,
So dawn goes down to day.
Nothing gold can stay.

-Robert Frost (1874–1963)[1]

The Red Wheelbarrow

so much depends
upon

a red wheel
barrow

glazed with rain
water

beside the white
chickens

-William Carlos Williams
(1883–1963)[2]

These poems and the poetry between them make me smile. I hadn't thought about their compatibility before, but both have a way of fending off the fantasy of permanence. Both invite us to pay attention, to mind whatever actual goodness comes around to visit. Look at this—the components and the composite and the fleeting beauty of it all. Taking in what a particular moment holds is a worthy project at any time. It feels especially so as I look out the window onto springtime in Vermont and try to receive what these sabbatical days have to offer.

I count your last letter among the gifts received lately. It arrived during my last week in the office, so it sat unopened for a

---

1. Frost, *Yale Review*, 30.

2. The poem now known as "The Red Wheelbarrow" was originally published with the title "XXII"; Williams, *Spring and All*, 74.

few days. But when I got to it, I was happy to be carried again by the current of our conversation. Your thoughts on forging partnerships and figuring out how to reach each other across the many divides, and finding "footbridge" decking in the stories we share—it all seems generously kind and invitational and necessary.

If I remember right, Katharine Hayhoe calls climate change an "everything issue." You and I have been talking about it as a "round problem." Same zip code, I think. Anyway, this human-made crisis does threaten everyone and everything, and it calls for a round response in word and action. So, the more people who gather around the table of climate change solutions the better. Our students in the Winter Term course taught me that. I was really struck by how those econ majors and psych majors and art majors and history majors and religion majors and environmental studies majors brought different questions, perspectives, assumptions, scars, hopes, and insights into the room. The expansiveness of those conversations was remarkable. I loved that.

So, yes, I'm all for actively pursuing partnerships, likely and unlikely ones. I also think it's important to check in with what each partner has to bring to this "everything issue." And even more, to hear from each one why climate change matters.

At church, our Green Team's newly minted Creation Justice Covenant speaks to what we hope to bring to our partnerships, near and far. Have you seen this?

> From the beginning, God entrusted us as stewards of God's good creation. In fashioning us in God's image, God bestowed upon us the responsibility to tend to creation as God would: with love and a respect for the needs of all living things. In affirming the divine gifts of creation and our connection to God, each other, and the world around us, the Congregational Church of Middlebury, UCC, commits ourselves to this urgent responsibility. We pledge to increase our awareness of how the abuses of creation cause environmental degradation and human suffering, and to celebrate and support work that preserves or restores ecological processes that benefit all life. We promise to fight the injustices of climate change, supporting and advocating for those most harmed by its

April 28, 2021

effects. Furthermore, as we confront this growing crisis, we resolve that these deeply felt commitments will be reflected in all dimensions of our congregation's life and extend far beyond our church's walls. In this way, we acknowledge and honor God's glory and perfect intent.[3]

I also find myself wanting to lean in to hear what other individuals and groups contribute to the "stone soup" work before us. What priorities and cares and knowledge and wisdom do each of us bring to ecological repair and environmental justice?

Why does climate change matter? What would you say?

I've been mulling it over. And even though I'm sure I'll want another whack at this, I can start here.

As a Christian, as a father, as an active member of my community, and as a human being on this planet, climate change matters to me, because loving God and loving my neighbor as myself includes responsibility for the well-being of the great web of life.

I believe that any attempt to love God includes love for the most vulnerable neighbors, humans and non-humans alike. What matters to me, in the marrow of me, is the kind of justice Israel's prophets demand and Christ embodies. I believe we love God by loving our neighbors and the whole neighborhood. And I believe we have an obligation under God to care for the poor and powerless and to speak out against the political, social, and economic circumstances that exacerbate the effects of climate change for the most vulnerable among us.

Does this mean that I don't think God's big enough to take on climate change? Nope. Does this mean that I'm a works-righteousness guy? Nope.

It means that as disciples of Christ, I believe we are endowed by God to be there for each other, especially in times of need.

Love God. Love your neighbor as yourself. That's what propels me in this work. And as the effects of climate change weigh down the world's most heavy-laden souls, inaction's hues of callousness, selfishness, greed, and ignorance stand in stark contrast to Christ's Great Commandment.

---

3. Congregational Church of Middlebury, "Our Creation Justice Covenant."

April 28, 2021

And again—I want to know what others have to say and what each has to bring to the table. I'm listening.

A few years ago, a friend shared with me an expansive story of sorts—something astronaut Russell "Rusty" Schweickart said aboard the Apollo 9 mission. It reads like a devotional:

> And you identify with Houston and then you identify with Los Angeles and Phoenix and New Orleans. And the next thing you recognize in yourself is that you're identifying with North Africa. You look forward to that, you anticipate it, and there it is. . . . When you go around the Earth in an hour and a half, you begin to recognize that your identity is with that whole thing. And that makes a change.
>
> You look down there and you can't imagine how many borders and boundaries you cross, again and again and again, and you don't even see them. . . . And from where you see it, the thing is a whole, and it's so beautiful. You wish you could take a person in each hand, one from each side in the various conflicts, and say, "Look. Look at it from this perspective. Look at that. What's important?"[4]

What's important?

There are so many ways to answer that question day to day. But what happens when our manifold responses nudge us toward something we all have in common—"the whole thing"? We call this planet home, for as long as we get to be here. The golden hues will fade, but we are given this moment to act as gratefully as we feel for God's gift of life. The next moves are ours to make. God bless our steps.

I wish you well, Andi—blessings on the final weeks of the semester.

Peace,
Andy

---

4 Schweickart, "No Frames, No Boundaries."

MAY 6, 2021

Dear Andy,

By my count, you have reached day twenty of blessed sabbatical time. I imagine—I hope—that the unwinding has continued. That you're settling into a different rhythm of time. It makes me glad to hear of your wanderings into the springtime loveliness, not least of all because the colors of that day found their way into your letter, and from there came to life in my imagination. I haven't yet found a forest in New Haven that quite replicates that first-green-is-gold hue of a Vermont maple forest in late April.

As I write this, I am back at home in the Green Mountains. I arrived here two nights ago after a stop on the way home to bring my dad Sally's pizza. It brought him much joy. Me, too. The light was fading as I crossed over Bethel Gap. That first glimpse of the east-facing ridge of the Green Mountains as you crest the gap is usually the moment that my heart sings out "home!" But as I descended into Rochester, I realized that I hadn't even noticed the mountains. My attention was fully focused on the conversation that awaited me at the end of the drive.

I wonder if we'll look back, years hence, and see the pandemic as a time of clear-sightedness. When time slowed enough and our worlds contracted enough that we could see clearly the cracks and the brokenness. Systemic brokenness, to be sure—the brokenness of our health care system, the brokenness of our political system, the health inequities produced by systemic racism. But also, the cracks in our own lives. The ones that we've chosen not to look at as we rush through our busy days. The ones that we've papered over, again and again. I know people who, as the pandemic unfolded, noticed the fault lines in their relationships, and healed them. People for whom time together was a needed balm. That hasn't been my

May 6, 2021

story, as you know. For Tom and me, time together in this year of pandemic isolation turned those fault lines into canyons.

Time away was the balm I needed to get to a place of peace and clarity. Clarity about what needed to change. Clarity that I wanted it to change. Clarity about all that I love about this life we've built on this little patch of land in the mountains. Clarity about all that I love about the person with whom I've built that life. And clarity that things weren't likely to end in that longed-for place called happily-ever-after, together. That combination—wanting it to be better and knowing it probably couldn't be—was heavy. I said a quick prayer in the dark of the driveway, then walked in the door of my home.

Tom met me at the door. We talked late into the night. It was honest. Calm. And so very hard. So very sorrowful. We told each other what was true about our lives. About ourselves as people. And we both listened. And we both knew before we said it out loud. This place we were standing in? It's the end of the road. Tom was the one to call the question. "So, we're calling it quits?"

I asked if we could wait a day before speaking the answer to that question out loud. Some sentences are best said by morning light. And also, I guess I just wanted to let all of the hard things that we had said rest awhile, there in the space between us. One more night of hoping that somehow our mutual pain might form a bridge. I fell into bed, too exhausted to sleep. I just lay there, listening to the lullaby of the wood frogs and peepers in our over-full pond. The peace of wild things. Life, happening.

∞

I'm writing this two days later. We've now spoken the answer to that question aloud. Yes. This is the end. It's not what I had hoped for. And yet, I can also see that my hopes weren't dashed. Not exactly. We reached this place with far more mutuality than I could ever have hoped, with far more love than I could ever have hoped. Love that knew it couldn't ask the other to become something less than whole. And so, this breaking apart has left me feeling strangely and surprisingly whole. And yes, also, of course: grief-struck, exhausted, hurt, humbled, with occasional swerves into the swampy

May 6, 2021

valley of shame. But whole. The integrity of who I am—and of who I am becoming—intact. More intact than before.

As the reality of all of that settles in, that's the word on my heart, perched there on top of those sighs too deep for words. Integrity. And as it turns out, it's the word to describe both my inner state and the answer to the question you asked me about climate change in your letter.

What matters to me about climate change? Integrity. That's what matters. What I want for this world is simple: wholeness. And climate change threatens that wholeness. That's why climate change matters to me.

I want a world that nurtures each of us into wholeness. A world in which each of us can be complete, can be who we are called by God to be. I think that's another way of saying the same thing you're pointing to when you talk about justice—about becoming a world in which we are here for each other, especially the most vulnerable. Climate change threatens to add millions—or billions—more people to the category of "most vulnerable." It's a threat to the wholeness of humanity in general and individual humans in particular. And that's not right.

And as an ecologist, a lover of this earth, and a human being who depends on the gifts God's creation provides, I want wholeness not just for individual people and the communities they form, but for the world. All of it. And climate change surely threatens the integrity of the world. Not just its people, but all of non-human creation.

I was thinking about this as I was writing my sermon for this coming Sunday—my last in Hinesburg—on that passage in John 15 in which Jesus commands the disciples to love one another as he has loved them. I read that and my first thought was: mycorrhizal networks. (This was before I saw that the *New York Times* wrote a story about Suzanne Simard's book *Finding the Mother Tree*. But I'm ordering that book today.)

When I had the thought, I was thinking about one of my favorite forests. It's an old-growth hemlock forest on the western slopes of the Green Mountains in Middlebury. No trails lead there. No signs announce its presence. You would never notice it from the hiking trail that runs past it. But it is, by some estimates, the

May 6, 2021

largest patch of uncut forest in Vermont: one hundred acres out of the ~1,000 acres of old-growth forest left in this state. It's mostly hemlock, with some red pines on the drier, more exposed slopes. As best as I can tell, the oldest trees are close to four hundred years old. One of these days, we'll take a walk up there. I think you'd like it.

From aboveground, the forest looks like a collection of individual trees. I've stood there with college students dozens of times over the years and explained that this is a forest shaped by competition. Rugged individuals competing for scarce resources, each staking out its claim to a patch of land. It turns out I was not entirely right about that. Although I'm sure that competition is part of what happens in the forest, the world of ecology has been moving toward a different paradigm these last several years. It's a paradigm that becomes clear if you look underground.

Underground, those individual trees are woven together. Into a community. I'd been missing the forest for the trees. Underground, a network of mycorrhizal fungi connects roots of neighboring trees. Trees share resources across that network. Resources flow from source to sink—from those with plenty to those in need. Canopy trees share with seedlings, and that helps those seedlings grow into their full potential. Trees communicate across that network, too. Alarm chemicals warn neighbors of insect attacks.

It feels to me a little bit like the kind of community I hear Jesus calling the disciples to form. A community of mutual caring. Of sharing what we have so that everyone has enough. A community that knows that our individual integrity is bound up in the integrity and well-being of the whole. Jesus tells the disciples to bind themselves together with other-centered love, one to another, and all—through Christ—to God. Be like a forest, I hear. Parts woven into a beautiful whole.

And climate change threatens all that. In my beloved old-growth hemlock forest, that threat has a name: hemlock woolly adelgid. It's an invasive insect—a lot like an aphid—that was accidentally introduced to Virginia from Japan in the 1950s. It's spread since then up and down the Appalachians. Those tiny insects, which look like flecks of cotton, can kill a hemlock tree in a handful of years. Whole forests disappear.

May 6, 2021

Did you ever spend time at the Reservoir in West Hartford when you were growing up? The one off Farmington Avenue? I spent a lot of time there as a kid, hiking. (And a lot of time there as a teenager, doing less wholesome things.) One of the trails I used to walk passes through a gorgeous hemlock forest. Dense canopy, wide open understory. The kind of forest that really does feel like a cathedral. I went back a few years ago and was saddened to find that the hemlocks are gone. Some of them were killed by adelgids, and some were cut down in a desperate attempt to slow the insect's spread.

Hemlock woolly adelgids have done their worst damage in more southerly climes. They don't overwinter well in the kind of cold we get in northern New England. They've been confirmed in Vermont, but they haven't wreaked the kind of havoc here as elsewhere. Not yet, at any rate. By the middle of this century, it's likely to be warm enough in this part of Vermont for hemlock woolly adelgid to thrive.[1] In which case, that community of interconnected hemlock trees is doomed. And all of the denizens of that forest—white-tailed deer, flying foxes, pileated woodpeckers, bobcats—will be displaced. Some of them will find new homes. Others won't.

The world will be less whole. Incomplete.

And that's happening all over the planet. Species extinctions. Range shifts. Changes in the physical environment—flooding, fire intensity. All contributing to the unraveling of the ecological fabric of this world. Eroding its integrity. That's why I care about climate change. And sure, it's partly pragmatic: our lives depend on the services and resources provided by intact ecosystems. But it's not just pragmatism. It's also love.

I delight in this world. In its beauty, its strangeness, its mystery. I love this world. I'm in love with it. And so, I want wholeness for it, as I want wholeness for everyone I love. This astonishing place, this Earth on which our lives depend, this creation that God named "very good" is at risk. It's at risk because of us. Our lives, as we are living them, imperil all life on Earth. As you said, the next moves are ours to make.

---

1. Paradis et al., "Role of Winter Temperature and Climate Change," 541–54.

May 6, 2021

A couple of years ago, I had to write a prayer of invocation for worship. Words weren't coming, so I went out into the backyard. I just sat in the dark and listened, eyes closed, to the sounds of life humming around me. Eventually words appeared.

> Gracious God, may we find peace in the quiet of you.
> May we find courage in the nearness of you.
> May we find wonder in the vastness of you.
> May we find faith in the mystery of you.
> Amen.

Peace, courage, wonder, and faith. May they guide our next moves.

Peace, friend,
Andi

MAY 19, 2021

Dear Andi,

Grace and peace.

It was good to see you last week. That climb to the four-hundred-year-old hemlocks, and to the red pines, was like stepping up and onto your front porch. There was a look of home about you among those trees. And then there was the walk back down along Abbey Brook—Vermont really does show off sometimes, doesn't it?

More than this, I'm glad we could find some time to walk and talk. The news you shared in your last letter, along with updates on the trail, were hard to hear and also lined with grace notes. The vulnerability and clarity with which you speak about the end of your and Tom's marriage is tender terrain. It brings to mind a passage from "Beyond Endings" by John O'Donohue:

> Creatures made of clay with porous skins and porous minds are quite incapable of the hermetic sealing that the strategy of closure seems to imply. The word completion is a truer word. . . . Though such an ending may be awkward and painful, there is a sense of wholesomeness and authenticity about it. Then the heart will gradually find that this stage has run its course and the ending is substantial and true. Eventually the person emerges with a deeper sense of freedom, certainty, and integration.[1]

God bless you. Bless your family, too.

After our walk through the old-growth forest, I sat in the shade of the flowering apple tree in our yard. I wondered how old it was. I've wondered this before, lots of times, but this time I wondered just enough to stand up, find a tape measure, and wrap it around its waist. The circumference of the stem is 96 inches; according to a

---

1. O'Donohue, *To Bless the Space Between Us*, 157.

tree age calculator I found online, that means our apple tree is about 103 years old.

Google's algorithms hard at work, I also learned that the oldest apple tree in the States died last year in Vancouver, Washington, at the age of 194.[2] Also, that the Great Bristlecone Pines in the White Mountains of California are pushing five thousand years old.[3] Five thousand? How am I just hearing this?

Anyway, come May, this old-for-an-apple-tree apple tree hums with bees. And it's roomy enough for a whole conservatory of birdsong. No wonder the psalmist includes trees—fruit trees and all cedars—in the chorus of creation's praise (Psalm 148:9, NRSV). When the apple blossoms bloom out front, words like "idyllic" and "Edenic" come to mind. I'm pretty sure I think more about the Garden of Eden in May than in all the other months combined. I have St. Jerome to thank for this—Eden's apple starts with him.

As you know, in the original Hebrew text, in Genesis chapter 3, the Tree of the Knowledge of Good and Evil bears "fruit." Just fruit. Not apples. Then, about twelve hundred years later, when Jerome translates Genesis 3 (along with most of the Bible) into Latin, he goes with "apple." It's wordplay. He chooses *malus* for "fruit," because malus can mean "evil" and "apple." So, Jerome drops a pun into Paradise. I love that.

The idea of "original sin" made its debut in the fourth century CE, back in Jerome's day. What are we to make of it in ours? That we are made in God's image, but we are not God? That we are inherently flawed?

As I hear it, the story of Adam and Eve eating fruit that they were asked by God not to eat is a holy reminder that there's some unholy mess in all of us. That's not the whole story, of course. In the opening chapter of Genesis, we are also "very good." But the leitmotif of human screw-ups is hard to miss, and in the shadows of Eden, it begins with disobedience (literally, "a refusal to listen"). That's the first sin of Hebrew Scripture—disobedience—followed quickly by an unflattering propensity to shirk responsibility and to

---

2. National Park Service, "Requiem for the 1820s Fort Vancouver Apple Tree."

3. National Park Service, "Bristlecone Pines."

May 19, 2021

blame the other. Adam and Eve hear God's instructions and choose not to listen.

When it comes to climate change, there is an ignominious legacy of not listening. As far as I can tell, not listening includes choosing to tune into voices that tell us that a warming planet is fake news and has little to do with human activity. Not listening includes listening to voices that privilege the self (and profits) above all else. Not listening includes selective hearing. I think we all know something about that.

Wasn't it Arrhenius who connected the theoretical dots between fossil fuel combustion and global warming? Wasn't that in the late 1800s? You know those details much better than I. But I think it's fair to say that Big Coal and Big Oil have known for well over a century that there was a very good chance that we'd end up where we are today.

In any case, *I* have known about climate change for years, but I still pay to heat our home with oil and schlep around in a gas-powered minivan. I've pressed my ear to the scientific data and, even more, to the wisdom of God's Word. I have been teaching and preaching stewardship of creation, and neighbor-love, and special care for the most vulnerable among us for nearly half my life. I still don't get it right. Maybe that's why Paul's admission to the Christians in Rome has a familiar ring to it: "What I don't understand about myself is that I decide one way, but then I act another, doing things I absolutely despise."[4]

If the story ended there, I'm not entirely sure I could throw my feet onto the floor each morning. But sin—*original* or otherwise—doesn't get the last word. The lostness of human disobedience is led home by grace. Grace—the unearned, unmerited, unconditional love of God—is greater than the sum of our sins. When Paul writes to the Romans that "nothing will be able to separate us from the love of God in Christ Jesus our Lord," (Romans 8:39, NRSV) he's talking about grace.

Grace is a holy kind of kindness. It's our way of talking about a God that loves anyway and always. "Nobody's perfect" is

---

4. Peterson, *The Message*.

May 19, 2021

demonstrably true. And yet, God wants us to flourish. I remember running into grace not long after returning to church in my mid-twenties. It must have been December '95. I was at Battell Chapel for an Advent service of evening prayer. I sat in a small circle of chairs in the chancel, and someone asked me to read a passage from Isaiah 61, the part about God wanting to give us a garland instead of ashes, the oil of gladness instead of mourning. As I got into the reading, I felt the tears coming.

I was a little raw going in. A couple of people I loved were falling apart. And the first semester finals were bearing down. So, maybe I was in the sweet spot to hear the good news that God wanted good for us, all of us, even me.

Anyway, grace got through to me, and it was more than my eyes could hold. I felt broken open and embarrassed. I didn't really know the people there. I hadn't been going to church there (or anywhere) for very long. They must have thought I was a real prize, but the saints of the Church of Christ in Yale responded to God's grace by offering a garland of life-giving kindness to the new guy who wasn't at full strength.

There are countless channels of grace bringing life to hard and hopeless places. And I've come to believe that such grace gives us the help we need to do what needs doing.

Now, I know some church-going folks who think—as Pelagius did in the fourth century—that we are perfectly able to live in alignment with God without divine aid. And I guess there's a part of me that wants to believe that, too. It's an alluring thought. But I know myself well enough to be suspicious of it. In fact, I'm reminded daily that I need God's help to nudge (or yank) me beyond the small, small world of I, me, mine.

As the woes of climate change accelerate, these reminders grow louder. *God, help me* and *God, help us* are prayers I pray a lot. They are prayers born of the belief that God can help. And does help. And will help again. I don't think God will don a red cape and swoop in to save the day. I do think that God finds countless ways to equip us for the tasks before us. And I think it's our work to use our God-given tools to help build beloved community, over and

May 19, 2021

over again. I also think that at the farthest limits of what we can contribute to this project, God's love is just finding its stride.

Is grace a blanket that shields us from the realities of a warming planet? No. But it does offer the assurance that we are not alone in the struggles of our day and the ones brewing on the horizon. By grace, God is with us. That's why, despite the unholy mess in my bones, I work and rest in hope.

I wish you well as you begin your CPE training, my friend. If your experience is anything like mine, it will stay with you for a long time.

I'm off to Eastern Point Retreat House soon for eight days of silence. It's time to listen.

With the oil of gladness,
Andy

JUNE 4, 2021

Dear Andy,

I was glad to get your letter—and to see that we share a tendency to form close, personal attachments to trees. That made me smile. I was glad, too, for the words of grace. Because who couldn't use a reminder of grace? Of the hope that springs from knowing the depth and breadth of God's love? And thank you, too, for those words from John O'Donohue. I'd been feeling the inadequacy of the language we have to describe the ending of relationships. Words like broken, failed. And there's truth in those. Things were broken. And I could write you a list of my failures a mile or two long. But that's not the whole of the story. So, I appreciate O'Donohue's contribution to the lexicon of endings. His words ring true for me. And I'm grateful to you, friend, for sending them along. It helps, you know. To feel heard. Less marooned on my own island of loss.

That recognition—of what it feels like to be met in a place of sorrow with understanding—was a pretty perfect realization to carry into the start of a summer of clinical pastoral education. I'm two weeks in and feeling every bit of my beginner status. I love the work—the being with people, praying with people, listening to people. It feels like a gift to spend my days with veterans. It's also one of the hardest things I've ever done. As you know far better than I, to be present to abyssal depths of pain and sorrow is to kneel down on some hard and holy ground. And I'm still trying to figure out exactly how to do that. I feel over my head much of the time. The question rings through my mind: what do I have to offer? But I'm trying hard to remember that as inadequate as I may feel in any given moment, my adequacy (or lack thereof) is not even close to the whole story. My notes from earlier this week reflect that learning. "Remember God," I wrote.

Remember God. It's good advice.

JUNE 4, 2021

    I've taken to going on a long walk in the late afternoon after I get home from the hospital. I need that time to reflect. To process. And ultimately, to let things go. And regardless of where I intend to go, my path bends toward the trails that run from the Eli Whitney Center into East Rock. It's a beautiful, tangled riparian forest. I go slow, trying to identify the trees. My memory of Connecticut trees is rusty, and I don't have a field guide. But the count of species is getting up there: at least three oak species, sugar maple, maybe silver maple, a couple of birches, a hickory or two, two species of cherry, and some magnificent elder beech trees. Oh, and the mountain laurel's in bloom in the understory. It's glorious there in the deep, restful, green light of late afternoon.

    Your letter put the word *Edenic* in my head—it feels exactly right for that stretch of forest. Despite the signs of human use—an old pipe crossing the stream, the trail, a bench, the hum of traffic—it feels *Edenic*. The tangled green gives the place a primeval feel. So, I came home from my forest wanderings the other day with that word on my lips and decided that a field trip into biblical Hebrew was warranted. You probably know where the name Eden comes from. But I didn't. And the answer is such a delicious linguistic tidbit that if you hadn't been on a silent retreat, I would have called you to share what I found. My Hebrew lexicon speculates that the proper name Eden derives from an obscure word that means, among other things, delights. Luxuries, dainties, delights.[1] So, Eden: the Garden of Delights.

    That's where I go in the late afternoon to shed the heaviness of the day's work: to the life-sustaining green, the source of the very air we breathe. And in the quiet of that enveloping green light? I remember our life-sustaining God, the source of the very lives we live.

    Jonathan Edwards, who surely walked those woods while at Yale three centuries ago, would say that that's grace at work. I've been spending quality time with Edwards in the little moments of free time these past couple of weeks; my professors from my spring history class on Edwards and American Puritanism have suggested that I consider revising my final paper to submit to a journal for

---

1. Brown et al., *Hebrew and English Lexicon*, 727.

June 4, 2021

publication. So, I'm considering. The paper was an effort to think through what Edwards's theology has to say to a modern Christian ecological ethic. Edwards was no environmentalist, not in any modern sense of the word. But I was interested in his ideas about why God created the world and what that vision of God has to say to us, the creatures made in God's image. The ones placed here to take care of this good Earth.

Edwards wrote a long essay called "The End for Which God Created the World,"[2] in which he set out to answer that question: why did God create the world? His answer is simple, surprising, and lovely. God created the world as an act of communication. God wanted to communicate God's own goodness—to spread it beyond God's own self—and so God created all that is. Edwards concluded that it was an act of communication spurred by pure delight: God's overflowing delight in God's own goodness. He puts it like this: "'tis fit, that he should take delight in his own excellencies' being seen, acknowledged, esteemed, and delighted in."[3]

Out of delight, God created the Garden of Delights. And here we are, gardeners, I suppose, of delight. Meant to reflect God's goodness, God's joy, God's delight back to God. That's what Edwards saw in the act of creation. As one who believes in God as Creator, I feel like I have to take that insight seriously.

I also feel like I have to take a short detour to explain that sentence—for my own sake. I remember the question coming up in our class. Maybe you do, too. What do I, with my PhD in evolutionary biology, think about a creating God? I remember stumbling over the answer that day. So, here's my second attempt, one more year into divinity school. Nowadays, what I mean when I say I believe in a creator God is that I believe this universe would not exist without God. And I mean that I believe God caused the universe to be, and God's creative involvement in the world is ongoing. I couldn't begin to explain how God's work of creation relates to evolution, except that I believe both to be true. I feel like every time somebody tries to build that bridge, it comes up short. So, I'm learning to befriend

---

2. Edwards, "The End for Which God Created the World," 406–536.
3. Edwards, "The End for Which God Created the World," 433.

June 4, 2021

the Mystery of it. But all of that is to say that it seems to me that whatever insight we humans can glean into God's relationship with creation is meaningful. Knowing something about why God created the world tells us something important about the world and about how we are to be in the world.

So I've been walking around wondering what it means that the world exists as a delight-fueled communication of God's goodness. What's in that insight for us? Especially now? Edwards argued that, for one thing, what's in it for us is the beautiful idea that we can read nature to learn about God. That's at least partly what motivated Edwards to become the exceptionally talented naturalist that he was. The seventeen-year-old author of a scientific publication on spider dispersal was doing theology, out there with the spiders.[4]

Edwards believed that we develop the capacity to do that—to sense the divine in the world—by grace. Grace gives us a spiritual sense, one that goes beyond where reason can take us. "There is a difference," he writes, "between having a rational judgment that honey is sweet and having a sense of its sweetness."[5] Same with God. By grace, we don't merely know or believe in the "loveliness and beauty" of God's holiness and goodness: we sense it. And we sense it, in part, out here in the world. The world that can be read as a book, one that tells the story of God's goodness.

Edwards gets positively poetic, describing what grace can equip us to see in the ordinary world all around us. This is from a short essay on "The Excellency of Christ," in which Edwards is describing what a spiritual sense allows one to see about Christ out there in the world:

> So the green trees and fields, and singing of birds, are the emanations of his infinite joy and benignity; the easiness and naturalness of trees and vines [are] shadows of his infinite beauty and loveliness; the crystal rivers and murmuring streams have the footsteps of his sweet grace and bounty. When we behold the light and brightness of the sun, the golden edges of an evening cloud, or the beauteous bow, we behold the adumbrations of his glory

4. Edwards, "The Spider Letter."
5. Edwards, "A Divine and Supernatural Light," 105–24.

> and goodness; and the blue skies, of his mildness and gentleness. There are also many things wherein we may behold his awful majesty: in the sun in his strength, in comets, in thunder, in the towering thunder clouds, in ragged rocks and the brows of mountains. That beauteous light with which the world is filled in a clear day is a lively shadow of his spotless holiness and happiness, and delight in communicating himself.[6]

You said it in your last letter: grace gives us the help we need. And seeing the world as a reflection of God's glory and goodness? An act of delighted, divine communication? A Garden of Delights? Surely, in this era of climate change, there is help to be found in that kind of awareness.

Surely, we would live differently on the earth if we sensed—really sensed, deep in our hearts, in the way Edwards describes—this world as a communication of God's own self. A reflection of God's own self. The product of God's own delight. What a different view of the world that is. We've gotten so accustomed to seeing the world in purely utilitarian terms: as a resource to harvest or use, as an object to control. But to see the world instead as a place reflecting and pulsing with God's own delight? A "world charged with the grandeur of God,"[7] as Gerard Manley Hopkins put it? Such a world is to be walked on with reverence. Barefoot, on holy ground.

And such a world is to be delighted in. To be loved. Seeing the world that way, I think we'd find it an awful lot harder to read the day's headlines—epic droughts in California and Taiwan, an extra-early start to the hurricane season, right whales teetering on the brink of extinction—and not be moved to act. And yes, that seems like a big ask. We don't even succeed at loving our human neighbors most of the time. Your letter describes our lostness well on that front. And also, grace. Grace that might just equip us to see this world differently. To see it as sacred. Not an object to be controlled, but a web of relationships to be nurtured.

---

6. Edwards, "Excellency of Christ."
7. Hopkins, *Poems*, 26.

June 4, 2021

That's the hope I find myself resting in, reading Edwards. That we can, by grace, learn to see differently. Learn to fall in love with this world. To love it, divine image-bearers that we are, as God loves it.

Edwards also leaves me with a question. I fell in love with this world as an ecologist. Summer after summer of kneeling before ancient trees, and my delight drew me in. But all this? It makes me wonder if maybe, without knowing it, by falling in love with the world, I was falling in love with God. Did the forests lead me home to God, delight drawing me in?

Maybe that's what my ecologist's eyes have been showing me all along. I just didn't fully recognize what I was seeing. That in the fabric that binds us all together—life, interconnected, interdependent—I was catching a glimpse of the threads of divine care that hold us. That in the green trees, the invisible, vibrating whir of photosynthesis, giving us food to eat and air to breathe, I was glimpsing God's bountiful care. That in the wild places and uncontrolled forces of nature, I was catching sight of a God who is untamed, wild. And that when my ecologist's eyes told me to walk on this good Earth gently, with reverence, they were seeing holy ground. Edwards would say that I needed grace to see the traces of God in this world, and it is surely by grace that I remembered God and began to see again what was so clear to me as a child: God with us. In all of this.

So, it's occurred to me lately that I might not actually be starting over. There's a lot I'm leaving behind on this journey, yes. But I'm suddenly seeing all of the things that I'll carry with me, too. I'm seeing that maybe ecologist and pastor aren't as far apart as I thought they were. They're different stretches of trail, yes, but it's all the same beautiful path through this one precious life. I've been caught up in God's holy ecosystem of delight the whole time—I just didn't know it.

Amazing grace. Because of which I, too, rest and work in hope, my friend.

I trust that you and yours are well, and that Vermont is resplendent as always at this turn of the seasons. And I hope that there was good listening to be had over those eight days of silence.

June 4, 2021

I look forward to hearing about it. In the meantime, enjoy the unfolding of these sabbatical days.

Grace and Peace,
Andi

JUNE 23, 2021

Dear Andi,

Greetings, on this refreshingly cool and sunny morning. We surely need rain here, more rain than yesterday's promising clouds delivered, but I'd rather not wish away this pale blue sky and its cotton ball clouds. Not yet, anyway. It's too good. And the family's fine, which adds joy to the full-fledged concert of birdsong outside my window. Mary's still on crutches from her trail-run-gone-bad, but she's hoping that her appointment tomorrow with the orthopedist will signal progress. Last Sunday, she insisted on hopping around the kitchen and baking me a strawberry rhubarb crisp for Father's Day. So delicious!

Around that time, I saw the photo you posted on social media—the one of you and a brother and a nephew and *your dad* on a Connecticut beach. Given what you've all been through this year, the joy emanating from the picture was not lost on me. I imagine that moment is still with you, as you make rounds at the hospital.

Your description of the first two weeks of clinical pastoral education is now two weeks old. I wonder how your words—"beginner status" and "over my head" and "inadequate"—sound now, a month or so into your summer stint at the VA. I suppose pastoral ministry bears these markings in perpetuity; I still feel the pangs of inadequacy after two dozen years. But your willingness to kneel on the "hard and holy ground" of others' suffering, and your note to self to "remember God," and your practice of letting things go on those laureled paths tell me that you *are* figuring it out. And then some. It sounds like the figuring that gives shape to a pastor's heart. Continued blessings on your ministry there, friend.

So, I need to say—your last letter is a bountiful feast. I'm still metabolizing it, still nourished by it. Thank you for that. I especially appreciate the generous serving of Jonathan Edwards's faithful

JUNE 23, 2021

imagination. The letter arrived not long after I returned from Eastern Point Retreat House, so Edwards's words, and yours, have been helping me make sense of that time in Gloucester. *My* figuring is a work in progress, of course, but here's a start.

The cold open brought me headlong into re-creation.

My first night there, a violent wind played whale songs up a downspout outside my room. That wind got all-mighty, like the One who works with wind was personally involved. I was lying awake for hours, ferried between delight and dread, wondering what would be made of me.

The wind lessened throughout the week, but it never left. God's breath kept sweeping over the depths of prayer and rest and walks on the rocks. In turns, the cession of words anchored me in holy peace and sent me to where the dragons dwell. "Notice what you notice," my eight-day spiritual director said.

Some of what I noticed was the stuff of me that needs to be extinguished. I'll spare you that. But I also noticed the rafts of Mallards and American Black Ducks and Common Eiders that call Brace Cove home. Actually, I sought them out. And when I found them, I watched them paddle-boat across the quiet water of Niles Pond and dive for dinner along the coast. (Is there such a thing as "benign stalking"?) The longer I watched, the happier I felt. God was getting through to me, the way Edwards and you make clear. It wasn't subtle either—in fact, it got a little silly. One afternoon, I walked a path to the big rocks to pray. I stood out there in the windblown rain. My eyes were open. The ocean was open. I didn't know what to say. So, I just listened and listened. At one point, the words "Come, Holy Spirit" popped out of my mouth, and right then, a female Mallard winged past me, very close, and quacked.

Seriously, God?

So it was, daily, in that patch of the Garden of Delights.

Weeks later, you and Edwards have me thinking more about the ways God communicates to and through all kinds of neighbors and friends, feathered and otherwise. I keep smiling at the thought of God finding "delight in his own excellencies being seen, acknowledged, esteemed, and delighted in." This kind of communication sounds like communion to me. In fact, one of the definitions of

June 23, 2021

"communion" *is* "communication," the kind of communication that manifests "intimate fellowship or rapport." Maybe we are brought into communion with the subjects of this world and with the Author of Life. Then again, maybe God's communication reveals a fellowship that's been there all along.

At a talk he gave in Calcutta in October 1968, Thomas Merton pointed to a holy unity that exists and persists, despite our mortal and moral failures of recognition. He opens up an apophatic mystery that I think I have sensed, if fleetingly, from time to time. He writes:

> The deepest level of communication is not communication, but communion. It is wordless. It is beyond words. And it is beyond speech. . . . My dear brothers [and sisters], we are already one. But we imagine that we are not. And what we have to recover is our original unity. What we have to be is what we are.[1]

As I hear it, Merton's insight that "we are already one" taxis us on the runway of ecological and theological truths. It's a space like the one your Earth Day Communion liturgy occupies. I've been meaning to tell you how much I appreciate the chords strummed in your Invitation to the Table:

> Our interdependence—with one another and with the rest of creation—is an ecological truth. We need each other. And we need the soil and the sun and the rain and the wind and our creaturely neighbors. And maybe it's the boundedness of our bodies, but we find it easy to forget that fact. To think of ourselves as unitary creatures rather than parts of a bigger whole. This meal—this sacrament of Holy Communion—invites us to remember otherwise. The very word, Communion, recalls us to the mutuality of our lives. To the truth that they are lived in relationship: with one another, with all of creation, with Christ. To the truth that it is in Christ that we find the fullest expression of the mutuality on which our lives rest.

Amen, sister.

1. Merton, *The Hidden Ground of Love*, x.

June 23, 2021

Communion, like communication, is something shared. Neither is a solo act. So, your invitation to "remember otherwise" at the Table calls to mind a deep sense of gratitude for what we all share in common—life on Earth, for starters.

Your words also help to center me in Christ, the One who shows and tells the nearness of God. The One who says, "love your enemies," and "do to others as you would have them do to you," and "I am with you always, to the end of the age." The One who says, "follow me," then blesses, breaks, and gives bread to whomever hungers. The One who sets my feet on a path toward flourishing life. A life in which reverence abides for God-made delights. A life in which we take Jesus *in* so that we can live Jesus *out* in manifold expressions of care, in remembrance of Him.

Communion helps these days, as we limp from the wreckage of a pandemic. Minds reeling from the longest year. Hearts rummaging for the muscle memory of social graces. I believe, deep down, that we are moving from what we've known to a remaking of life in community. And as we make our way, poet Lucille Clifton offers a good word, a benediction, a poem called "blessing the boats":

(at St. Mary's)

> may the tide
> that is entering even now
> the lip of our understanding
> carry you out
> beyond the face of fear
> may you kiss
> the wind then turn from it
> certain that it will
> love your back   may you
> open your eyes to water
> water waving forever
> and may you in your innocence
> sail through this to that[2]

I hope the onset of summer ushers in more barefoot moments for you in the company of veterans in West Haven, and on those

---

2. Clifton, *How to Carry Water: Selected Poems of Lucille Clifton*, 130.

June 23, 2021

familiar paths below East Rock, and in the ongoing reunions with family and friends. My family and I look forward to seeing you in Vermont next month. Until then, may the love of God, the peace of Christ, and the communion of the Holy Spirit be with you.

    With gratitude,
    Andy

P.S. On the eighth day, on the drive home from Eastern Point, the road *did* rise to meet me. The wind *was* at my back. Such is the regenerative power of God, the DNA of grace. Isn't it something?

JUNE 7, 2021

NOAA scientists reported today that in May, atmospheric CO_2 reached 419 ppm. This was the highest monthly reading ever recorded at the Mauna Loa Observatory.[1]

1. National Oceanic and Atmospheric Administration, "Carbon Dioxide Peaks near 420 Parts per Million."

Bibliography

350.org. "350 Campaign Update: Divestment." https://350.org/350-campaign-update-divestment/.

Barber, William. "William Barber's Inaugural Sermon as Joe Biden Takes Office." January 21, 2021. *Time Magazine*. https://time.com/5931343/william-barber-inaugural-prayer-service-sermon/.

Barnes, Michele L., et al. "Social Determinants of Adaptive and Transformative Responses to Climate Change." *Nature Climate Change* 10 (2020) 823–28.

Barnette, Martha and Grant Barrett. "Episode 1557," *A Way With Words*, National Public Radio, November 14, 2020.

Bartlett, David. *Westminster Bible Commentaries: Romans*. Louisville: Westminster John Knox, 1995.

Beers, F.W. *Atlas of Addison Co. Vermont*. New York: F.W. Beers & Co., 1871.

Berry, Wendell. "The Idea of a Local Economy." In *The Art of the Commonplace: The Agrarian Essays of Wendell Berry*, edited by Norman Wirzba, 249–61. Berkeley: Counterpoint, 2002.

———. *Life Is a Miracle: An Essay Against Modern Superstition*. Berkeley: Counterpoint, 2000.

Biden, Joseph R. "Inaugural Address by President Joseph R. Biden, Jr.," January 20, 2021. https://www.whitehouse.gov/briefing-room/speeches-remarks/2021/01/20/inaugural-address-by-president-joseph-r-biden-jr/.

Braun, Mike. "S.1251—117th Congress (2021-2022): Growing Climate Solutions Act of 2021." Congress.Gov. June 24, 2021. https://www.congress.gov/bill/117th-congress/senate-bill/1251.

Briggs, Reinette et al., eds. *Principles for Building Resilience: Sustaining Ecosystem Services in Social-Ecological Systems*. Cambridge: Cambridge University Press, 2015.

Brown, Francis, et al. *Brown-Driver-Briggs Hebrew and English Lexicon*. Snowball, 2010.

Brueggemann, Walter. *Journey to the Common Good*. Louisville: Westminster John Knox, 2010.

Bullard, Robert D., et al. "Toxic Wastes and Race at Twenty: Why Race Still Matters After All of These Years." *Environmental Law* 38 (2008) 371–411.

BIBLIOGRAPHY

Byers, Edward, et al. "Global Exposure and Vulnerability to Multi-Sector Development and Climate Change Hotspots." *Environmental Research Letters* 13 (2018) 055012. https://doi.org/10.1088/1748-9326/aabf45.

Cambridge University. "Early Records," January 28, 2013. https://www.cam.ac.uk/about-the-university/history/early-records.

Chambers, Suzanna. "At Last, a Degree of Honour for 900 Cambridge Women." The Independent. May 30, 1998. https://www.independent.co.uk/news/at-last-a-degree-of-honour-for-900-cambridge-women-1157056.html.

Chapin, F. Stuart. "Triggering Transformation to Sustainability through Stewardship." November 15, 2019. Stockholm Resilience Centre, Stockholm, YouTube, 47:56. https://www.youtube.com/watch?v=4iSqF45XOTQ.

Charles, Dan. "Farmers Are Warming Up To The Fight Against Climate Change." National Public Radio. November 20, 2020. https://www.npr.org/2020/11/20/936603967/farmers-are-warming-up-to-the-fight-against-climate-change.

Clements, Frederic E. "Nature and Structure of the Climax." *Journal of Ecology* 24 (1936) 252–84.

Clifton, Lucille. *How To Carry Water: Selected Poems of Lucille Clifton*. Rochester: BOA Editions, Ltd., 1991.

Climate Analytics and NewClimate Institute. "Climate Action Tracker." Accessed July 11, 2021. https://climateactiontracker.org/.

Climate and Clean Air Coalition. "Mexico." Accessed July 11, 2021. https://www.ccacoalition.org/en/partners/mexico.

Commission for Racial Justice. *Toxic Wastes and Race In the United States: A National Report on the Racial and Socio-Economic Characteristics of Communities with Hazardous Waste Sites*. Cleveland: United Church of Christ, 1987.

Common Dreams. "Cambridge University Makes Historic Break With the Fossil Fuel Industry and Commits to Full Divestment." October 1, 2020. https://www.commondreams.org/newswire/2020/10/01/cambridge-university-makes-historic-break-fossil-fuel-industry-and-commits-full.

Congregational Church of Middlebury. "Our Creation Justice Covenant." Accessed July 11, 2021. https://www.midducc.org/church-activities/green-team.

Davis, Ellen. "Meaning of Dominion." Bible Odyssey. Accessed July 11, 2021. https://www.bibleodyssey.org/en/passages/related-articles/meaning-of-dominion.

Douglas, Kelly Brown. "A Christian Call for Reparations." Sojourners, June 5, 2020. https://sojo.net/magazine/july-2020/christian-call-case-slavery-reparations-kelly-brown-douglas.

Ecological Society of America. *Preservation of Natural Conditions*. Springfield, IL: Ecological Society of America, 1921.

Edwards, Jonathan. "A Divine and Supernatural Light." In *A Jonathan Edwards Reader*, edited by John E. Smith, Harry S. Stout, and Kenneth P. Minkema, 105–24. New Haven: Yale University Press, 1995.

Bibliography

———. "The End for Which God Created the World." *Ethical Writings*, Writings of Jonathan Edwards Online. 1749. http://edwards.yale.edu/research/browse.

———."108. Excellency of Christ." *The "Miscellanies": (Entry Nos. a-z, Aa-Zz, 1–500)*, Works of Jonathan Edwards Online. 1722. http://edwards.yale.edu/research/misc-index.

———. "The Spider Letter." *Scientific and Philosophical Writings*, Writings of Jonathan Edwards Online. 1714. http://edwards.yale.edu/research/browse

Egerton, Frank N. "History of Ecological Sciences, Part 47: Ernst Haeckel's Ecology." *Bulletin of the Ecological Society of America* 94 (2013) 222–44.

Frost, Robert. "Three Poems." *The Yale Review* XIII (1923) 30.

Gayle, Damien. "Cambridge Accepts £6m Shell Donation for Oil Extraction Research," The Guardian. 11/5/2019. http://www.theguardian.com/education/2019/nov/05/cambridge-accepts-6m-shell-donation-for-oil-extraction-research.

Gould, Stephen Jay. "This View of Life: The Panda's Peculiar Thumb." *Natural History* 87 (1978) 20–30.

Harris, Ben. "Does RBG's Rosh Hashanah Death Really Make Her a 'Tzadik'?" The Times of Israel, September 22, 2020. https://www.timesofisrael.com/does-rbgs-rosh-hashanah-death-really-make-her-a-tzadik/

Hopkins, Gerard Manley. *Poems of Gerard Manley Hopkins*, edited by Robert Bridges. London: Humphrey Milford, 1918.

Intergovernmental Panel on Climate Change. *Climate Change 2014: Synthesis Report. Contribution of Working Groups I, II and III to the Fifth Assessment Report of the Intergovernmental Panel on Climate Change*. Geneva: IPCC, 2014.

International Renewable Energy Agency, "SIDS—Small Island Developing States." Accessed July 11, 2021. https://islands.irena.org/.

Jefferson, Dawn. "When Miriam Is Missing," sermon, Marquand Chapel at Yale Divinity School, New Haven, CT, September 28, 2020.

Joerstad, Mari. *The Hebrew Bible and Environmental Ethics: Humans, Nonhumans, and the Living Landscape*. Cambridge: Cambridge University Press, 2019.

Jordan, Rob. "Soil Holds Potential to Slow Global Warming." Stanford News, October 5, 2017. https://news.stanford.edu/2017/10/05/soil-holds-potential-slow-global-warming/.

Kahan, Dan M., et al. "Culture and Identity-Protective Cognition: Explaining the White-Male Effect in Risk Perception." *Journal of Empirical Legal Studies* 4 (2007) 465–505.

Lee, Charles, Ed. *Proceedings: The First National People of Color Environmental Leadership Summit, Principles of Environmental Justice, October 1991*. New York: United Church of Christ Commission for Racial Justice (now Justice and Witness Ministries, a Covenanted Ministry of the United Church of Christ, Cleveland, Ohio). Accessed 7/11/2021 from https://www.ucc.org/what-we-do/justice-local-church-ministries/justice/faithful-action-ministries/environmental-justice/principles_of_environmental_justice/.

BIBLIOGRAPHY

Leopold, Aldo. *A Sand County Almanac with Essays on Conservation From Round River.* New York: Ballantine, 1966.

Lopez, Barry. *Arctic Dreams: Imagination and Desire in a Northern Landscape.* New York: Charles Scribner's Sons, 1986.

Matthews, H. Damon. "Quantifying Historical Carbon and Climate Debts among Nations." *Nature Climate Change* 6 (2016) 60–64.

Memmott, Mark. "'Trillions Of Earths' Could Be Orbiting 300 Sextillion Stars." National Public Radio. December 1, 2010. https://www.npr.org/sections/thetwo-way/2010/12/01/131730552/-trillions-of-earths-could-be-orbiting-300-sextillion-stars.

Meredith, Sam. "Oil Major Total's Full-Year Profit Falls 66% as Covid Pandemic Hits Fuel Demand." CNBC, February 9, 2021. https://www.cnbc.com/2021/02/09/total-earnings-q4-full-year-2020.html.

Merton, Thomas. *The Hidden Ground of Love: The Letters of Thomas Merton on Religious Experience and Social Concerns.* Edited by William H. Shannon. New York: Farrar, Straus, Giroux, 1985.

Mockenhaupt, Brian 2015 "Why Being a Good Neighbor Is a Good Idea," *High Country News.* December 7, 2015. https://www.hcn.org/issues/47.21/why-being-a-good-neighbor-is-a-good-idea.

Monks of the Weston Priory, "Go Up to the Mountain," The Benedictine Foundation of the State of Vermont, Inc., 1978.

National Park Service. "Bristlecone Pines—Great Basin National Park (U.S. National Park Service)." Accessed July 11, 2021. https://www.nps.gov/grba/planyourvisit/identifying-bristlecone-pines.htm.

———. "Requiem for the 1820s Fort Vancouver Apple Tree, and a New Dawn (U.S. National Park Service)." Accessed July 11, 2021. https://www.nps.gov/articles/000/requiem-for-1820s-fort-vancouver-apple-tree.htm.

National Oceanic and Atmospheric Administration. "Carbon Dioxide Peaks near 420 Parts per Million at Mauna Loa Observatory—Welcome to NOAA Research." NOAA Research News. June 7, 2021. https://research.noaa.gov/article/ArtMID/587/ArticleID/2764/Coronavirus-response-barely-slows-rising-carbon-dioxide.

National Oceanic and Atmospheric Administration. "Rise of Carbon Dioxide Unabated—Welcome to NOAA Research." NOAA Research News. June 4, 2020. https://research.noaa.gov/article/ArtMID/587/ArticleID/2636/Rise-of-carbon-dioxide-unabated.

New Oxford American Dictionary, s.v. "self-interest." New York: Oxford University Press, 2010. iOS version published by Mobi Systems, Inc.

O'Donohue, John. *To Bless the Space Between Us.* New York: Doubleday, 2008.

OED Online, s.v., "Respair, n," https://www-oed-com/view/Entry/275567

———. s.v., "Respair, v," https://www-oed-com/view/Entry/163776

———. s.v., "Respire," https://www-oed-com/view/Entry/163821

Oxfam International. "5 Natural Disasters That Beg for Climate Action." https://www.oxfam.org/en/5-natural-disasters-beg-climate-action.

BIBLIOGRAPHY

Olson, Mark E., et al. "A User's Guide to Metaphors In Ecology and Evolution." *Trends in Ecology & Evolution* 34 (2019) 605–15.

Palmer, Parker. "Parker Palmer on Standing in the Tragic Gap." Center for Courage & Renewal, August 21, 2013. http://www.couragerenewal.org/723/.

Paradis, Annie, et al. "Role of Winter Temperature and Climate Change on the Survival and Future Range Expansion of the Hemlock Woolly Adelgid (Adelges Tsugae) in Eastern North America." *Mitigation and Adaptation Strategies for Global Change* 13 (2008) 541–54.

Peterson, Eugene, trans. *The Message: The Bible in Contemporary Language*. Colorado Springs: NavPress, 2014.

Pope Francis. *Encyclical on Climate Change & Inequality: On Care for Our Common Home*. Brooklyn: Melville House, 2015.

Rauschenbusch, Walter. *Prayers of the Social Awakening*. Boston: Pilgrim, 1910.

Reubold, Todd. "Katharine Hayhoe: Bridging the Climate Change Divide." Ensia. June 16, 2016. https://ensia.com/interviews/katharine-hayhoe-bridging-the-climate-change-divide/.

Robinson, Robert. "Come, Thou Fount of Every Blessing." In *Chalice Hymnal*, 836. St. Louis: Chalice, 1995.

Schoon, Michael L., et al. "Principle 7—Promote Polycentric Climate Governance Systems." In *Principles for Building Resilience: Sustaining Ecosystem Services in Social-Ecological Systems*, edited by Reinette Biggs, Maja Schlüter, and Michael L. Schoon, 226–50. Cambridge: Cambridge University Press, 2015.

Schweickart, Russell. "No Frames, No Boundaries," Context Institute. September 16, 2011. https://www.context.org/iclib/ic03/schweick/.

Sittler, Joseph. "Nature and Grace in Romans 8." In *Evocations of Grace: Writings on Ecology, Theology and Ethics*, edited by Steven Bouma-Prediger and Peter Bakken, 207–22. Grand Rapids: William B. Eerdmans, 2000.

Solnit, Rebbeca. "On Not Meeting Nazis Halfway." Literary Hub. November 19, 2020. https://lithub.com/rebecca-solnit-on-not-meeting-nazis-halfway/.

Tansley, A. G. "The Use and Abuse of Vegetational Concepts and Terms." *Ecology* 16 (1935) 284–307.

Taylor, Matthew. "Cambridge University to Divest from Fossil Fuels by 2030," The Guardian. October 1, 2020. http://www.theguardian.com/education/2020/oct/01/cambridge-university-divest-fossil-fuels-2030-climate.

———. "Cambridge University Urged Again to End Fossil Fuel Investments." The Guardian. April 23, 2018. https://www.theguardian.com/education/2018/apr/23/cambridge-university-urged-again-to-end-fossil-fuel-investments.

United Church of Christ. "A Movement Is Born: Environmental Justice and the UCC." https://www.ucc.org/what-we-do/justice-local-church-ministries/justice/faithful-action-ministries/environmental-justice/a_movement_is_born_environmental_justice_and_the_ucc/.

Westminster Assembly. *Westminster Shorter Catechism*, 1647.

Williams, William Carlos. *Spring and All*. Paris: Contact Publishing, 1923.

Bibliography

Yale Program on Climate Change Communication. "Global Warming's Six Americas." Accessed July 11, 2021. http://climatecommunication.yale.edu/about/projects/global-warmings-six-americas/.

———. "Yale Climate Opinion Maps 2020." Accessed July 8, 2021. https://climatecommunication.yale.edu/visualizations-data/ycom-us/.

Young Evangelicals for Climate Action. "YECA supports Environmental Justice for All Act." Accessed July 11, 2021. https://yecaction.org/blog/overview.html/article/2021/03/22/yeca-supports-environmental-justice-for-all-act.

www.ingramcontent.com/pod-product-compliance
Lightning Source LLC
Chambersburg PA
CBHW070915160426
43193CB00011B/1473